The Guide for Manufacture, Installation, and Design of FRP Flanges and Gaskets

Printed and Bound in the United States of America

ISBN 978-1-57698-045-3

Published by Materials Technology Institute, Inc.
www.mti-global.org

Table of Contents

Preface

As part of its ongoing mission to fill the materials and technological "voids" which challenge the Chemical Process Industry, members of the Materials Technology Institute (MTI) sought to consolidate industry best practices and capture the "state of the art" as relates to fiber reinforced plastic flanges, the compilation of which are presented in this publication.

The MTI Resource Group for Project 160-04 that guided this effort, consisted of the following individuals:

Juan Bustillos (Dow Chemical)
Terry Cowley (DuPont)
Dale Heffner (Electrochemical Engineering & Mfg. Company)
Bob Hendrix (Eastman Chemical Company)
David Jensen (Eli Lilly & Company)
Dale Keeler (Dow Chemical)
Pradip Khaladkar (DuPont)
E. L. (Gene) Liening (Dow Chemical)
Edward Naylor (Akzo Nobel Chemicals Inc.)
Bernhard Rijpkema (Akzo Nobel Chemicals Inc.)
W. B. A. (Sandy) Sharp (MeadWestvaco Corporation)
Robert Sinko (Eastman Chemical Company)
Bob Smallwood (Det Norske Veritas)
Hardin Wells (Albemarle Corporation)

Their comments, guidance, and support were instrumental in our success with this project and appreciated.

While this document started out as an "outsourced" expert project, over time it became apparent that the true expertise and experience resided within the MTI membership. An "old school" approach was adopted with a few select members taking charge of authoring and editing the document. Special acknowledgement in this case goes to these principal authors, who contributed significantly to the scope and content:

Juan Bustillos (Dow Chemical)
Dale Keeler (Dow Chemical)
Hardin Wells (Albemarle Corporation)

Abstract

This guidebook provides an accessible, comprehensive review of current procedures and best practices for the design, manufacture, installation, and maintenance of FRP flanges. The information herein was compiled from various standards, technical publications, engineering handbooks, and industry sources. Each chapter in this guidebook discusses different aspects of FRP flange connections, including the design of flanges, the design procedures outlined in widely used industry standards, some of the most common manufacturing techniques, gaskets, and bolt-up procedures.

1
Introduction

Over the last several decades, the use of fiber-reinforced plastic (FRP) materials in piping systems has been increasing due to their corrosion resistance, low maintenance cost, and relatively long life expectancy. Although these attributes offer distinct advantages over traditional metal materials, FRP materials are more complex than metals; consequently, design and installation of FRP structures is more involved. As FRP materials for piping systems increase in popularity, procedures, and guidelines for the design and construction of components, such as FRP flanges, continue to develop.

The purpose of this guidebook is to provide an accessible, comprehensive review of current procedures and best practices for the design, manufacture, installation, and maintenance of FRP flanges. The information herein has been compiled from various standards, technical publications, engineering handbooks, and industry sources. Each chapter in this guidebook discusses a different aspect of an FRP flange connection. Chapter 2 covers the design of flanges, and discusses the design procedures outlined in several widely used industry standards. Chapter 3 discusses some of the most common manufacturing techniques, including the advantages and disadvantages of each. Chapter 4 discusses the proper use of gaskets and Chapter 5 covers bolt-up procedures for FRP flanges. In each of the chapters, some issues or "pitfalls" to be aware of are highlighted. Finally, example design calculations are given in Appendices A, B, and C.

Supplemental discussions on flange and gasket issues can be found in:

Guidelines for Safe Seal Usage – Flanges & Gaskets
the European Sealing Association and the Fluid Sealing Association (jointly)
Publication No. 009/98
September 1998.

Industrial Gaskets
Jose Carlos Veiga
Teadit N.A.
ISBN 09643120409

Process Equipment Design
Brownell & Young
Wiley
ISBN 0471113190

Formulas for Stress and Strain 3rd. Ed.
Table XIII Case 26
Raymond J. Roark
McGraw-Hill

Mechanical Design of Heat Exchangers and Pressure Vessel Components
Singh & Soler
Arcturus
ISBN 0916877019

2
FRP Flange Design

This chapter addresses the design of FRP flanges. The two major flange types are presented, followed by design methods for both types. This discussion includes various standards frequently used for design. The chapter then provides a brief example design calculation and ends with a section on Issues to Consider.

Flange Types

There are two types of flanges that are commonly used in FRP piping systems, tanks, and vessels: full-faced flanges and lap-joint flanges.

A full-faced flange consists of a flat, circular face connected to a tapered hub, with bolt holes integrated directly into the face (Figure 2-1). A lap-joint flange consists of a flat, circular face connected to a tapered hub and is clamped to an adjacent flange with a metallic or FRP backing ring that contains the bolt holes (Figure 2-2). Because the flange face does not contain bolt holes, the FRP face diameter for lap-joint flanges is smaller than that of full-faced flanges for a given inside diameter. In either case, gaskets used with FRP flanges always cover the full FRP face to avoid detrimental bending stresses in the FRP during bolt-up.

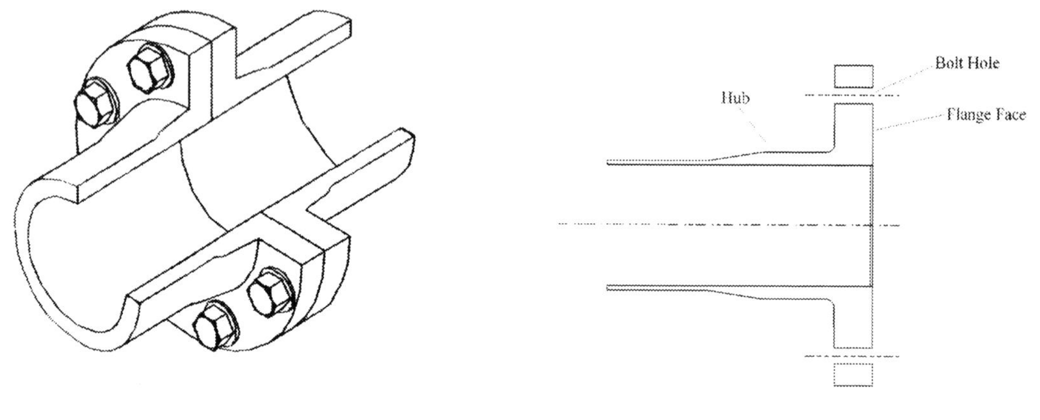

Figure 2-1. Schematic of a Full-faced Flange.

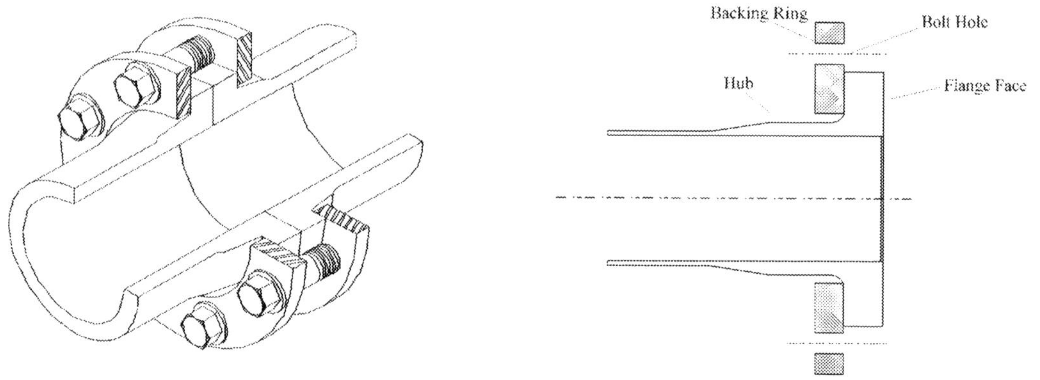

Figure 2-2. Schematic of a Lap-Joint Flange.

Advantages/Disadvantages

Lap-joint flanges have several advantages over full-faced flanges. First, because the diameter of the flange face is smaller, there is less area over which the bolt force is distributed onto the gasket. This translates into appreciably higher compressive stresses on the gasket material. It has been shown that this higher concentration of force on the gasket results in greater leak resistance.[1] In addition, the bending moments at the flange face/hub interface are smaller for lap-joint flanges because the clamping force of the bolts (via the backing ring) is concentrated closer to the hub, plus the bending moment in the backing ring is decoupled from the FRP flange face and primarily reduced to an axial force. Therefore, the bolts in lap-joint flanges can generally be torqued to a greater extent than for full-faced flanges without damaging the flange. The potential for increased bolt force is especially helpful in protecting against leaks when gaskets made of materials with lower elasticity and higher stiffness (such as PTFE based) are used in lieu of elastomers. Finally, the backing ring that clamps a lap-joint flange can rotate, which facilitates the process of lining up the bolt holes during installation.

Although lap-joint flanges have these advantages, full-faced flanges are used more extensively in North America. A potential explanation is that most standardized FRP flange design methods are derived from methods used to design full-faced metallic flanges.

Consequently, there is a greater knowledge base regarding the design of full-faced FRP flanges. In addition, in corrosive environments, an unprotected metal backing ring will be the "weakest link" in the system in terms of corrosion resistance. Replacing a corroded backing ring requires cutting the pipe and installing another ring or installing a split lap ring. For this reason, FRP or encapsulated metal backing rings are sometimes used. However, FRP backing rings are not as tolerant as metallic rings to crushing damage from the bolts. Finally, for large diameter flanges, the ease of bolt hole alignment during the installation of lap-joint flanges may be offset by difficulties associated with handling the weight of the metal backing rings.

The relative advantages and disadvantages of the two flange types are summarized in Table 2-1:

Flange Type	Advantages	Disadvantages
Full-faced	• Large design and use knowledge base (most common design used)	• Reduced bolt load capability • Lower gasket forces/less tight seal • Adapters are normally required when connecting to non-flat-face components • More "drawback" potential during fabrication
Lap-joint	• Higher gasket forces/tighter seal • Potentially reduced leakage • Easier installation in smaller sizes • Smaller bolt-up bending moments • Compatible with raised face components like plastic-lined metal piping and valves	• Fewer fabricators are able to offer • Possible crushing of FRP backing ring • Large flanges have heavy backing rings • Analytical design methodology not provided in most construction standards

Table 2-1. Comparison of Full-faced and Lap-Joint Flanges.

Construction

By definition, FRP flanges are constructed of layers of fiber-reinforced thermoset plastic resin. The type of reinforcement and resin used in an FRP flange should be consistent with those used in the piping system, tank, or vessel in which the flange will be integrated. Glass fibers are the most popular reinforcement used in industry, but other reinforcements, including carbon and aramid fibers, are also used. Thermosetting plastic resins such as vinyl esters, polyesters, epoxies, furans, and phenolics are used extensively. Thermoplastic resins such as PVC, CPVC, polypropylene, and PVDF are seldom used in composite applications because they have appreciably lower mechanical strength, creep resistance, and thermal stability than thermoset resins; however, they are used successfully as liners in "dual laminate" systems. Dual laminate construction is defined as a thermoplastic layer (used as a liner) with a bonded FRP structure.

A typical FRP flange used in the transport of corrosive materials consists of a surface layer, an interior layer, and a structural layer (Figure 2-3). The surface and interior layers, collectively called the corrosion barrier or the corrosion liner, are the innermost layers of the flange and are intended to protect the outer layers from chemical attack.

Surface Layer

The surface layer is typically composed of one or two thin resin-rich veils of corrosion-resistant reinforcement material (e.g., C-glass or synthetic veil), and is on the order of 0.010 to 0.020 in (0.25 to 0.50 mm) thick.

Interior Layer

The interior layer typically contains one or two plies of resin-rich 1.5 oz/ft^2 (450 g/m^2) chopped strand mat (CSM), but can also be fabricated from multiple plies of 0.75 oz/ft^2 (225 g/m^2) CSM. The thickness of the corrosion barrier in a typical 2 veil and 2 [1.5 oz/ft^2 (450 g/m^2)] CSM system is usually on the order of 0.100 in (2.54 mm). Specific applications may require greater or fewer layers of CSM to produce corrosion barrier thicknesses anywhere from 0.050 in (1.27 mm) to 0.250 in (6.35 mm). The reinforcement content of the corrosion barrier is usually 20% to 30% by weight (veils plus CSM). Each application requires careful selection of the proper corrosion barrier thicknesses. As an example, C-glass veil is good for many applications but is not recommended for caustic or strong alkaline-type environments. The resin selection and catalyst systems also play a crucial role in the success of the application.

Structural Layer

The structural layer serves as the load-bearing component of the flange and consists of plies of resin reinforced with multiple plies of CSM, woven (bi-directional) roving, and/or continuous strand roving. The directionality of the fibers in these plies, along with the glass-to-resin ratio, determines the strength and stiffness of the flange. The thickness of the structural layer is determined by design, as discussed below. There are several methods of manufacturing FRP flanges, which will be discussed in Chapter 3.

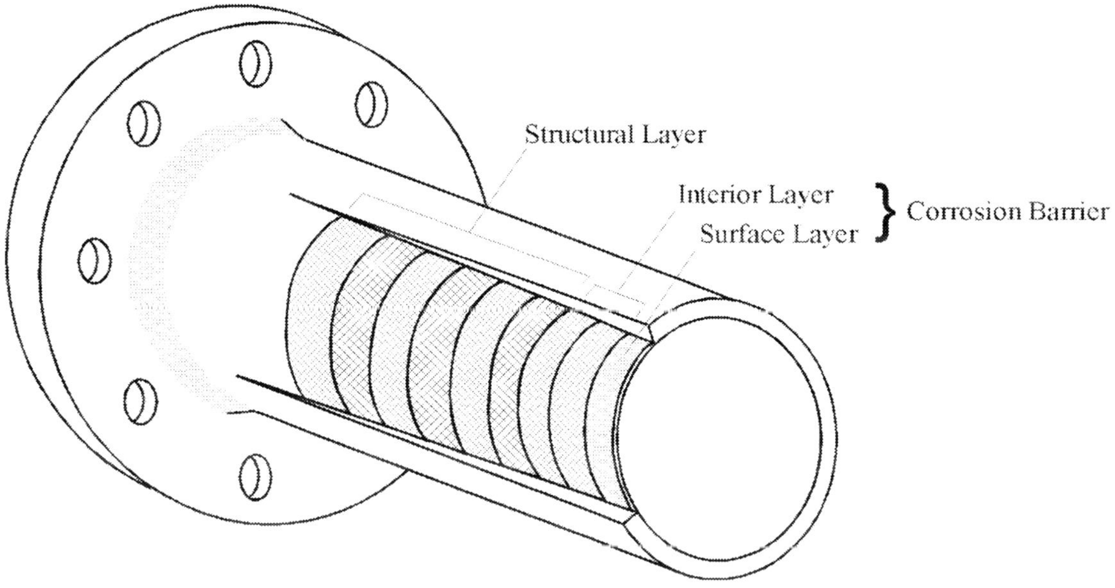

Figure 2-3. Typical FRP Flange Construction.

Design Methodology

There are several U.S. and European standards that describe methods of FRP flange design. Many manufacturers use one or more of these standards to design their FRP flanges, while others have either developed modifications to the standards or have developed their own methods.[III,IV,V] Flange manufacturers who have developed their own design methods can use ASTM D5421 (for contact molded flanges) or ASTM D4024 (for machine-made flanges) to qualify their designs. This section will review the flange design methods described in four widely-used FRP piping and vessel standards: National Bureau of Standards PS 15-69 (which is no longer in publication), ASME RTP-1, ASME Boiler & Pressure Vessel Code Section X, and British Standard 4994.

The flange itself is only part of a flanged connection, with the other important components being the gasket and bolts. To begin the design process using these standards, the designer must not only choose the flange type to be used but also an accompanying gasket material with known *m* and *y* factors (see Chapter 4) and a bolt material with known allowable tensile stress. The standards can then be used to complete the flange design by determining the following parameters (see Figure 2-4):

a) flange thickness

b) hub thickness

c) hub length

d) flange face diameter

e) bolt circle diameter (for full-faced flanges)

f) number of bolts

g) diameter of the bolts to be used

The design methods in the selected standards focus on calculating the flange thickness. All of the standards require that the hub thickness be at least one-half the flange thickness and taper uniformly down to the thickness of the host cylinder. In addition, the standards require that the hub length be at least three times the flange thickness. Items d through g usually match those of traditional metallic flanges, and the standards either provide tables of recommended values for these parameters or simply reference metallic flange standards (e.g., ANSI B16.1, ANSI B16.5, BS1560).

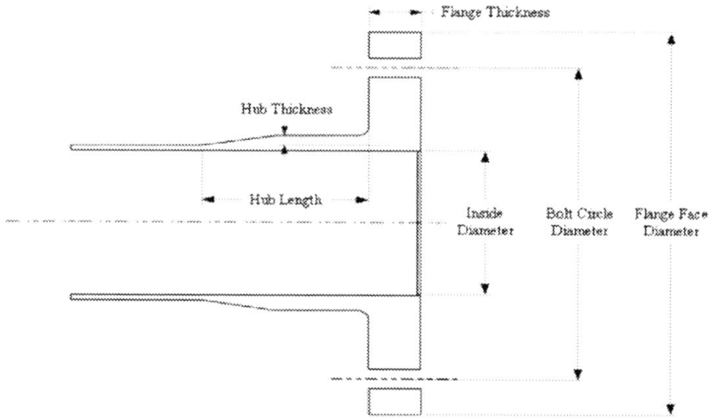

Figure 2-4. FRP Flange Dimensions.

Limitations of the Selected Standards

It is important to bear in mind that each of the selected standards has a particular range of applicability. The flange type, maximum design pressure, maximum operating temperature, materials, and manufacturing limitations for each standard are listed in Table 2-2.

Note that the U.S. standards include only design procedures for full-faced flanges, and all of the standards apply only to contact molded flanges (see Chapter 3).

Standard	Limitation				
	Flange Type	Max. Design Pressure Allowed for Flanges	Max. Temp	Materials	Construction
PS 15-69	Full-faced	150 psi (1.0 MPa)	180°F (82°C)	Polyester resin; glass fiber reinforcement	Contact molded
ASME RTP-1	Full-faced	Not Specified	180°F (82°C)*	Vinyl or polyester resin; glass fiber reinforcement, allows the use of thermoplastic liners. User specifies whether liners are to be included in the design.	Contact molded
ASME B&PVC Section X	Full-faced	Not Specified	250°F (110°C)	Epoxy, vinyl ester, polyester, furan resin; glass, carbon, aramid fiber reinforcement. Corrosion liners are excluded from design.	Contact molded
BS4994	Full-faced, Lap-joint	Not Specified	250°F (110°C)	Epoxy, polyester, furan resin; glass fiber reinforcement; allows use of thermoplastic corrosion liners.	Contact molded

* Higher temperature requires laminate tests at elevated temperature.

Table 2-2. Limitations of Selected FRP Piping and Vessel Standards.

Summary of Design Methods

The design methods and important specifications of each standard are summarized below. All pipe flanges with the exception of the duct and body flanges in the U.S. Standards and those in BS-4994 are based on class 150 B16.1 and B16.5 dimensions and bolting.

National Bureau of Standards Voluntary Product Standard PS 15-69 Custom Contact Molded Reinforced-Polyester Chemical Resistant Process Equipment 1969

PS 15-69, which covers FRP ducting, pipe, and tanks, was the first standard to offer design guidelines for FRP flanges. Although several more comprehensive and more modern standards have succeeded it, it is still used by some manufacturers in the construction of FRP flanges for piping systems and flanged FRP nozzles for tanks.

Laminate Composition Specifications

- Surface layer: A resin-rich surface layer 0.010 in (0.25 mm) to 0.020 in (0.51 mm) thick, reinforced with a glass veil, synthetic fibers or other material, as the application requires.
- Interior layer: A resin-rich layer at least 0.10 in (2.54 mm) thick, reinforced with non-continuous glass strands.
- Structural layer: A layer consisting of a combination of plies of resin reinforced with glass mat, cloth, or woven roving.

Design Chart

The flange thickness is given in a table for pipe sizes of 2 in. to 42 in. (50 mm to 1050 mm) and for pressure ratings up to 150 psi (1.0 MPa). The thickness values are based on a laminate flexural strength of 20,000 psi (138 MPa) and a safety factor of 8 to 1. The table also assumes the use of a soft full-faced gasket with Shore A or Shore A2 Hardness of 40 to 70. ANSI B16.1 and B16.5 are referenced for the flange face diameter, bolt circle diameter, number of bolts to be used, and the bolt diameter.

The standard provides tolerances for the flange face drawback and flatness, stating that the face should be perpendicular to the axis of the pipe within 1/2° and shall be flat to ±1/32 in (0.80 mm) up to and including 18-in (457 mm) inside diameter flanges and ±1/64 in (0.40 mm) for larger diameters.

ASME RTP-1
Reinforced Thermoset Plastic Corrosion Equipment
2000 Edition

RTP-1 covers the design, construction, fabrication, quality control, and inspection of reinforced thermoset plastic vessels containing corrosive or hazardous materials. The standard is limited to vessels operating at pressures not exceeding 15 psi (0.10 MPa) above any hydrostatic head and specifically excludes piping, rotating equipment, ducts, and stacks.

Laminate Composition Specifications

- Surface layer: A resin-rich surface layer 0.010 in to 0.020 in (0.25 mm to 0.5 mm) thick, reinforced with one or more veils made of fiberglass, carbon fiber, or organic fibers.
- Interior layer: A resin-rich layer with a nominal thickness of 0.10 in (2.54 mm), reinforced with chopped fiberglass mat or roving.
- Structural layer: Type I laminates (all mat) consisting of plies of 1.5 oz/ft^2 (450 g/m) fiberglass mat and Type II laminates (mat and roving) consisting of alternating plies of 1.5 oz/ft^2 (450 g/m^2) fiberglass mat and 24 oz/yd^2 (800 g/m^2) woven roving. For design purposes, Type I and Type II laminates are assumed to have maximum allowable design stresses of 1,800 psi (12.4 MPa) and 3,000 psi (20.7 MPa), respectively, reflecting a 5:1 design factor for the flanges.

- The user must make a choice as to whether or not the corrosion barrier can be included in the structural calculations.

Design Methods

RTP-1 offers two flange design methods, both described in non-mandatory appendices. The first method (Appendix NM-2 "Design of Integral Body Flanges") based on the Taylor Forge Method is intended for the design of full-faced flanges. This method allows the use of smaller bolts for body flanges but defaults to ANSI B16.1 and B16.5 for standard flanges. The second method (Appendix NM-12 "FRP Flange Design") is also geared toward the design of full-faced flanges but uses a different approach. RTP-1 also allows design by other techniques that determine the biaxial stress state in the flange, such as the finite element method, closed form solutions, back calculation of stress from strain gage data, etc. The standard does not describe these techniques in detail but provides a list of safety factors that should be followed to qualify a design if these methods are used. While there is no explicit limitation on flange design pressure, 50 psi (.34 MPa) is the "standard" design pressure outlined in Part 4 of Figure 4-6 of the document, although higher pressure designs are permitted by using the non-mandatory methods. The methods of the non-mandatory appendices are summarized below.

Method #1 (Non-Mandatory Appendix NM-2)

This procedure is based on the Taylor Forge method, which was developed over a half century ago for the design of full-faced metallic flanges. The procedure has two major steps. First, the designer identifies the flange face diameter, bolt circle diameter, number of bolts to be used, and the bolt diameter. A table is provided with values of these parameters for FRP body flanges with inside diameters ranging from 24 in to 144 in (609 mm to 3658 mm) and design pressures ranging from 5 to 30 psi (0.03 MPa to 0.21 MPa). Note that the standard recommends smaller bolts for the body flanges than the larger bolts specified in the ANSI metallic flange standards, which can sometimes cause problems with the FRP material because of space limitations;[6] in other words, since the body flanges in the document will only see very low pressures in comparison with a typical flange, the bolts are sized for that application rather than following the typical ANSI standard. Larger bolts will not only require thicker flanges but the larger diameter bolts will also require larger washers, which in some cases (because of the hub requirements) can be a problem with interference at the hub. Large washers require machining of the hub, thus reducing the hub strength because of the removal of material around the bolt hole.

The designer iterates through a series of design formulae to determine the thickness of the flange required to ensure that the radial and tangential stresses in the flange and the longitudinal stress in the hub are lower than the maximum allowable design stress for the laminate. These stresses are calculated by considering the bending moments in the flange caused by the required bolt load (for both gasket seating and operating conditions) and the pressure forces during operation.

Method #2 (Non-Mandatory Appendix NM-12)

This procedure (developed by A. D. Osborne for ASME RTP-1) is based on evaluating and controlling the bending stress in the flange during bolt-up and under operating conditions. Similar to Method #1, the designer must first identify the flange outside diameter, the bolt circle diameter, the number of bolts, and the bolt diameter. Guidance on these parameters is not explicitly provided, but an example calculation uses parameters comparable to those of an ANSI Class 150 weld neck flange. The designer then uses bending stress formulae to determine the thickness of the flange required to ensure that the flexural stress in the flange during both bolt-up and operating conditions is less than the laminate allowable stress.

The procedure also calculates the minimum bolt torque required based on seating and operating bolt loads as well as the maximum bolt torque allowed without damaging the flange or the bolts themselves. Finally, formulae are included for calculating the nozzle wall thickness, bolt spacing, and clearance.

The standard provides a flange face tolerance both for drawback and waviness and outlines the maximum allowed according to diameters in a table. (Table 4-1, Fig. 4-14 of NM-12)

ASME Boiler and Pressure Vessel Code Section X
Fiber-Reinforced Plastic Pressure Vessels
2003 Edition

ASME B&PVC Section X covers the design, construction, fabrication, testing, and inspection of fiber-reinforced plastic vessels. The standard describes two design classes covering vessel components such as flanges: Class I designs are based on the qualification of a design prototype through destructive testing (this is a method that lends itself to mass production of identical or similar vessels), while Class II designs are based on mandatory design rules and qualified through non-destructive testing.

Laminate Composition Specification

The standard does not place restrictions on the laminate composition, instead leaving the fabricator to both decide the laminate composition and use standard testing procedures to determine the mechanical properties (e.g., modulus, strength) of the laminate required for design. The standard allows for the incorporation of corrosion liners but states that the strength of the liner should not be included in design calculations.

Design Method

The standard describes a mandatory design procedure for Class II flanges, including flanged nozzles. This procedure is essentially the same as the Taylor Forge based procedure described in Method #1 of ASME RTP-1; however, it covers up to 150 psi (1 MPa) design, with caution to consider special design for larger diameter and high pressure applications. This is done because the typical flange dimensions do not allow enough distance to the bolt circle when the hub is fabricated to the flange thickness. The designer uses a series of formulae to calculate the flange thickness required to ensure that the stresses in the flange are less than the maximum allowable stress. The method specifies that the maximum allowable design stress be 0.001 times the lower of the longitudinal or transverse modulus of the laminate used to construct the flange. ANSI B16.1 and B16.5 are referenced for the flange outside diameter, the bolt circle diameter, the number of bolts, and the bolt diameter.

The standard provides a flange face tolerance, stating that face drawback and waviness shall not exceed 1/32 in (0.80 mm) as measured at or inside the bolt circle.

British Standard 4994
Design and Construction of Vessels and Tanks in Reinforced Plastics
1987 Edition

BS4994 covers the design, materials, construction, inspection, testing, and erection of reinforced plastic vessels and tanks. The standard describes three vessel categories (Category I, II, and III), which take into account factors including the contents to be carried, the chemical compatibility of the liner with the process fluid, design temperature and pressure, and size and geometry of the vessel.

Laminate Composition Specification

- Surface layer: For thermosets, a resin-rich surface layer 0.010 in (0.25 mm) to 0.020 in (0.5 mm) thick, reinforced with C-glass surface mat, synthetic fibers, or other suitable material. Thermoplastic lining with bond strength of 1015 psi (7 N/mm^2) in direct shear and 29 lb/in (5 N/mm) width in peel is also allowed.
- Interior layer: For thermosets, a layer reinforced with a minimum of 3.9 oz/ft^2 (1.2 kg/m^2) chopped glass strand or chopped strand mat with a soluble binder with 25% to 33% glass content by mass.
- Structural layer: No restriction on composition is specified. Guidance on mechanical properties of layers of resin reinforced with E-glass chopped strand mat, woven roving cloth, and unidirectional filament is provided. The standard indicates that laminates containing different types of reinforcement must be tested for mechanical properties.

Design Methods

BS4994 specifies design methods for both lap-joint flanges and full-faced flanges, and full face with backing ring flanges. For each flange type, the standard offers a table wherein the thickness of the flange face is specified for given values of design pressure and design strain of the laminate. The standard describes a procedure for determining the appropriate design strain of the laminate, which is based on either (a) the known fracture strain of the laminate divided by a safety factor of 10 or (b) the allowable unit loading, which is the ultimate tensile unit strength of the laminate divided by a safety factor that takes into account manufacturing procedure, chemical environment, design temperature, cyclic loading, and curing procedure. This safety factor can never be less than 8.

In addition, BS4994 offers a design procedure for full-faced flanges that is similar to the Taylor Forge based method of ASME RTP-1 and ASME B&PVC Section X, with two key differences. First, the flange thickness calculation is based only on ensuring the radial stress at the bolt circle is less than the laminate allowable stress. Furthermore, the laminate allowable stress is based on the aforementioned allowable unit loading and the weight per unit area of glass reinforcement.

The standard specifies that the flange face diameter, bolt circle diameter, number of bolts to be used, and bolt diameter should be in accordance with Class 150 of BS1569, Class 150 of BS3293, or Table 10 of BS4504.

Design Example

To compare the results of the selected standards, the thickness of an example flange was determined using each of the methods described above. The design was carried out assuming:

- A full-faced flange with inside diameter = 24 in (610 mm)
- Design pressure = 25 psi (0.17 MPa)
- A gasket with $m = 0.5$ and $y = 50$ psi (0.34 MPa)
- Bolt design tensile stress = 18,800 psi (130 MPa)
- Laminate mechanical properties:[1] ultimate tensile strength = 12.0×10^3 psi (82.7 MPa); tensile modulus = 1.3×10^6 psi (8,963 MPa)

In addition, the following flange parameters were used for each method:[2]

- Flange face diameter = 29.5 in (750 mm)
- Bolt circle diameter = 28 in (711 mm)
- Number of bolts = 20
- Bolt diameter = 0.5 in (12.7 mm)
- Bolt hole diameter = 0.625 in (15.9 mm)

For illustrative purposes, the step-by-step calculations using ASME RTP-1 Methods #1 and #2 are given in Appendices A and B, respectively. The results for all methods are given in Table 2-3. PS 15-69, ASME B&PVC Section X, and BS 4994 produce very similar flange thickness results, averaging approximately 1.18 in (30.0 mm) with a deviation of only 0.11 in (2.8 mm). The results of both methods of ASME RTP-1 are about 30% less than the other standards. This is because, for the chosen laminate, the allowable design stress assumed by RTP-1 (3,000 psi) has a less conservative safety factor than those used by the other standards.

Standard		Flange Thickness
NBS PS 15-69		1.13 in (28.7 mm)
ASME RTP-1[3]	Method #1	0.813 in (20.7 mm)
	Method #2	0.821 in (20.9 mm)
ASME B&PVC Section X[4]		1.30 (33.0 mm)
BS 4994[5]		1.10 in (27.9 mm)

Table 2-3. Design Example Results.

Some items of interest are as follows:

- ASME B&PVC Section X provides the most conservative design in this case, in part because it employs the lowest allowable flange design stress for this case (1,300 psi [9 MPa]).
- ASME RTP-1 Method #1, which is essentially the same procedure as ASME B&PVC Section X, provides a less conservative design due to its higher allowable design stress of 3,000 psi (21 MPa).
- BS 4994 employs an allowable design stress that is similar to ASME B&PVC Section X for this case (1,500 psi [10.5 MPa]). However, this method does not take into account stresses generated during gasket seating, and the use of the gasket (y = 50 psi [0.35 MPa]) in this example will result in higher stresses during gasket seating.

Issues to Consider

Flange Fracture versus Leakage

The design methods described in this chapter are all based on ensuring that the stress in the flange and the hub are less than the maximum allowable stress of the laminate, using conservative safety factors. As a result, these methods produce a flange that should not fracture under operating conditions. However, it is far more likely that a flanged connection will begin to leak before it ever fractures. To properly address the potential for leakage, a more sophisticated design method that includes the effects of deformations in the flange under operating conditions is required. Typically, such a method will employ finite element computer modeling of the flange and will require far more time and effort than the standard methods.

Current finite element models can explicitly include the various layers and their corresponding properties. Some uncertainty, however, remains, given issues like resin sitting between plies and any terminations of plies within the structure. These simulations are still very useful in understanding how stresses act within the flanges and allow the designer to anticipate stress concentrations and issues of gasket sealing reliability. These models can also simulate time-based relaxation of the FRP and gasket materials, providing an estimate of reliability over time.

Stress in the Hub

The designer should be aware that longitudinal stresses in the hub can be large. For instance, in the design example given above, ASTM RTP-1 Method #1 calculates a longitudinal hub stress of approximately 1,250 psi (8.6 MPa). Drawback (see next topic) and/or overtorqued bolts in such a flange could lead to stresses in the hub that exceed the allowable design stress. Drawback will often result in cracking at the hub-flange interface (Figure 2-5), while excessive applied stress levels can even result in catastrophic failures (Figure 2-6). Additionally, for higher pressure applications involving significant flange and hub thicknesses, the designer should consider the negative effects of material removed to accomodate the bolts, washers, and nuts.

Flange Face Drawback

As indicated above, flange face drawback (Figure 2-7) is an important issue. As will be discussed in Chapter 3, flange face drawback can be a problem in contact molded flanges. When a flange with excessive drawback is bolted up, large bending stresses can result.[VII] These stresses are not taken into account in any of the selected design methods, although PS 15-69 and ASME B&PVC Section X specify drawback tolerances. While the use of conservative safety factors in the design methods should take into account some stress related to drawback, the designer should be aware that excessive drawback might produce stress in the flange beyond that which was anticipated.

Stress in the area between the bolt circle and the outside diameter of the flange

The designer should be aware that some manufacturing techniques such as filament wound flanges sometimes do not offer the strength required to overcome some of the loading conditions. Very often because of dimension limitations dictated by standards such as ANSI B16.1 and B16.5, the area between the bolt circle and the outside diameter of an FRP flange is not strong enough to overcome the stress created by the bolts during bending, and this condition can result in cracking between the bolt holes and the flange OD (Figure 2-8). Careful consideration should be given to these types of fabrication.

Spot Facing

Spot facing the back of the flange surface provides a flat, parallel surface for the use of flat washers (Figure 2-9). By doing so, and utilizing flat washers, it is possible to minimize point loads on the back of the flange, and likewise provide more uniform loading of the flange by distributing the bolt forces across a larger area. Flat washers also prevent nut embedment into the back of the flange which could result in premature flange failure as well as large variations in the actual bolt force generated.

Sealing Ribs

Sealing ribs (or rings) consist of raised circular rings on the sealing surface of the flange face. These rings typically project no more than 1/32 in (.79 mm) from the flange face, have a radial width of around 1/16 in (1.58 mm), and may have a square, triangular, or rounded cross-sectional profile (Figure 2-10). For smaller flange sizes at least two or three rings are commonly used, while larger flanges may have appreciably more. The purpose of the sealing ribs is to create localized "peaked" or elevated compressive stresses on the gasket material thereby producing a tighter seal without causing detrimental stresses in the FRP flange. The ribs also increase the blowout resistance of a given gasket by creating a much higher frictional "bite" between the flange surfaces and the gasket. Sealing ribs that project out further than 1/32 in (.79 mm) may be used, but the risk of cutting softer gasket materials is much greater depending upon the bolt torques used. Furthermore, taller sealing ribs are more prone to mechanical damage or chipping during handling and assembly. Damage to the sealing ribs could in turn negate the benefit of the ribs, and in some cases actually create a leak path.

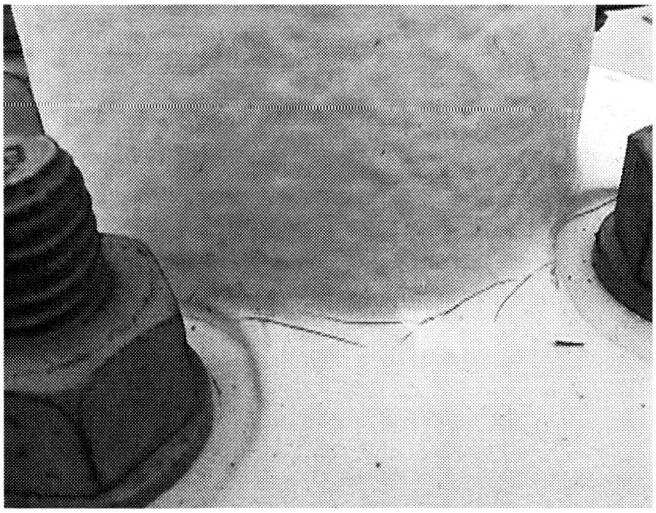

Source: The Dow Chemical Company

Figure 2-5. Cracks at the Flange-Hub Interface.

Source: Albemarle Corp.

Figure 2-6. Catastrophic Failure of Flange at Hub-to-Flange Intersection.

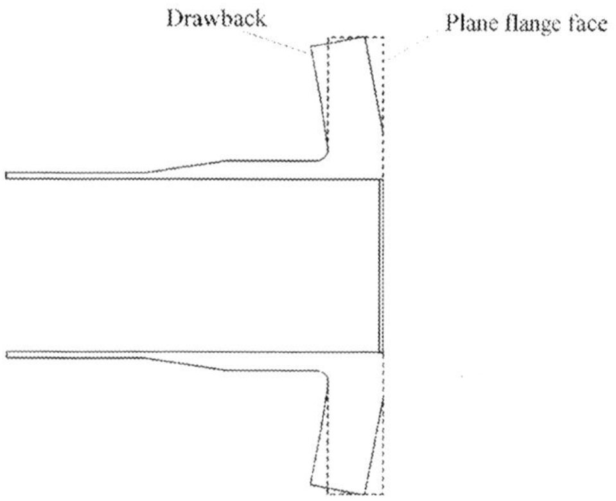

Figure 2-7. Cracks at the Flange-Hub Interface.

Figure 2-8. Flange Cracking (Bolt Circle – OD).

Source: Enduro Vesca Plastics

Figure 2-9. Spot Facing.

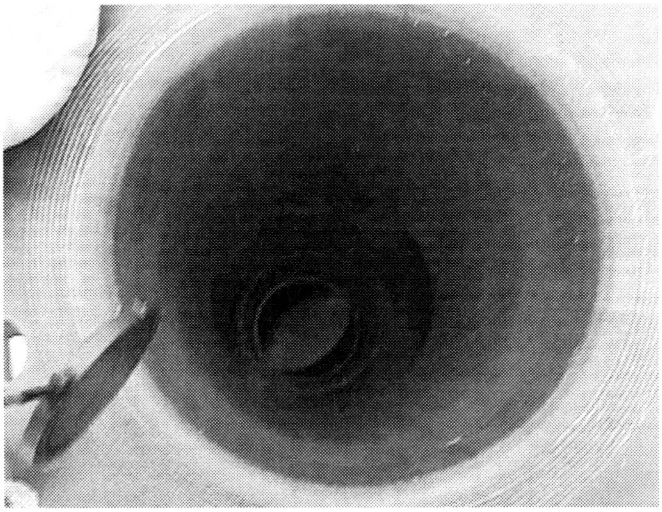

Source: The Dow Chemical Company

Figure 2-10. Sealing Ribs.

References

I. Godwin, E.W., Matthews, F.L., and Kitty, P.F., "The Design of Bolted Flanged Joints in GRP," Plast. Rubber Process. Applic., 1986, Vol. 6, pp. 161-167.

II. Sun, L., "Bolted Flanged Connections Made of Fiber Reinforced Plastic Materials," Ph.D. Dissertation, 1995, Concordia University, Montreal, Canada.

III. Private communication, Bill Bittner Sr., Bittner Industries.

IV. Private communication, Rick Watkins, Fibrex Corporation.

V. Private communication, Bob Newbold, RL Industries.

VI. Private communication, Juan Bustillos, The Dow Chemical Company.

VII. Blach, A.E. and Hoa, S.V., "The Effects of Pull-Back on Stresses in FRP Flanges," Exp. Tech., 1988, Vol. 12, pp. 12-16.

Bibliography

a. National Bureau of Standards, Voluntary Product Standard PS 15-69, Custom Contact-Molded Reinforced Polyester Chemical-Resistant Process Equipment, 1969.

b. ANSI/ASME B16.1 Gray Iron Pipe Flanges and Flanged Fittings: Classes 25, 125, and 250.

c. ANSI/ASME B16.5 Pipe Flanges and Flanged Fittings: NPS 1/2 through 24.

d. ASME, RTP-1-2000, Reinforced Thermoset Plastic Corrosion Resistant Equipment, 2001.

e. ASME, Boiler and Pressure Vessel Code Section X, Fiber-Reinforced Plastic Pressure Vessels, 2003.

f. ASTM D4024.

g. ASTM D5421.

h. British Standards Institution, BS 1560, British Standard Specification for Design and Construction of Vessels and Tanks in Reinforced Plastics, 1987.

i. British Standards Institution, BS 4994-87, British Standard Specification for Design and Construction of Vessels and Tanks in Reinforced Plastics, 1987.

j. British Standards Institution, BS 1569, Circular Flanges for Pipes, Valves and Fittings (Class Designated). Steel, Cast Iron and Copper Alloy Flanges. Specification for Cast Iron Flanges.

k. British Standards Institution, BS 3293, Specification for Carbon Steel Pipe Flanges (Over 24 Inches Nominal Size) for the Petroleum Industry.

l. British Standards Institution, BS 4504, Flanges and bolting for pipes, valves and fittings. Metric series.

1 Values taken from ASME RTP-1 Table 2A-3 for a 0.22 in thick Class II laminate
2 Values taken from ASME RTP-1 Table NM2-1
3 The maximum allowable design stress taken as 3,000 psi for the Class II laminate
4 The maximum allowable design stress assumed to be 0.001 x the tensile modulus: 0.001 x (1.3 x 106) = 1,300 psi
5 The minimum safety factor of 8 was used to determine the maximum allowable design stress from the ultimate tensile stress: 12,000 psi / 8 = 1,500 psi

3
FRP Flange Manufacture

Techniques commonly used to manufacture FRP flanges include contact molding (hand lay-up) and various automated procedures, such as compression molding, resin transfer molding, and filament winding. These techniques are described in the sections that follow, and the advantages and disadvantages of each technique are summarized in Table 3-1.

Contact Molding (Hand Lay-Up)

Contact molding is the simplest fabrication technique and involves applying layers of resin-impregnated mat or woven roving to a mold by hand until the desired thickness is reached (Figure 3-1). Saturation of the layers with resin is carried out during the process, and the resin is typically either sprayed or brushed on each layer after it is positioned on the mold. Curing of the resin can occur at room temperature or at elevated temperatures when heat lamps or hot air blowers are used. Elevated temperature curing, typically called post curing, produces a higher degree of cross-linking of the thermoset resin, along with maximum chemical resistance. Molds can be made of any material that holds a shape, and metals and plastics are commonly used.

Figure 3-1. Contact Molded FRP Flange Fabrication.

It is very important to fabricate these types of flanges with continuous fibers into the hub region. The use of reinforcing rings alone can create a very weak area at the flange to pipe intersection and a very weak hub, potentially resulting in a flange failure (Figure 3-2). Figures 3-3, 3-4, and 3-5 illustrate a suggested method for fabricating hand lay-up stub flanges.

Source: Albemarle Corp.

Figure 3-2. Catastrophic Failure of Flange at Hub-to-Flange Intersection.

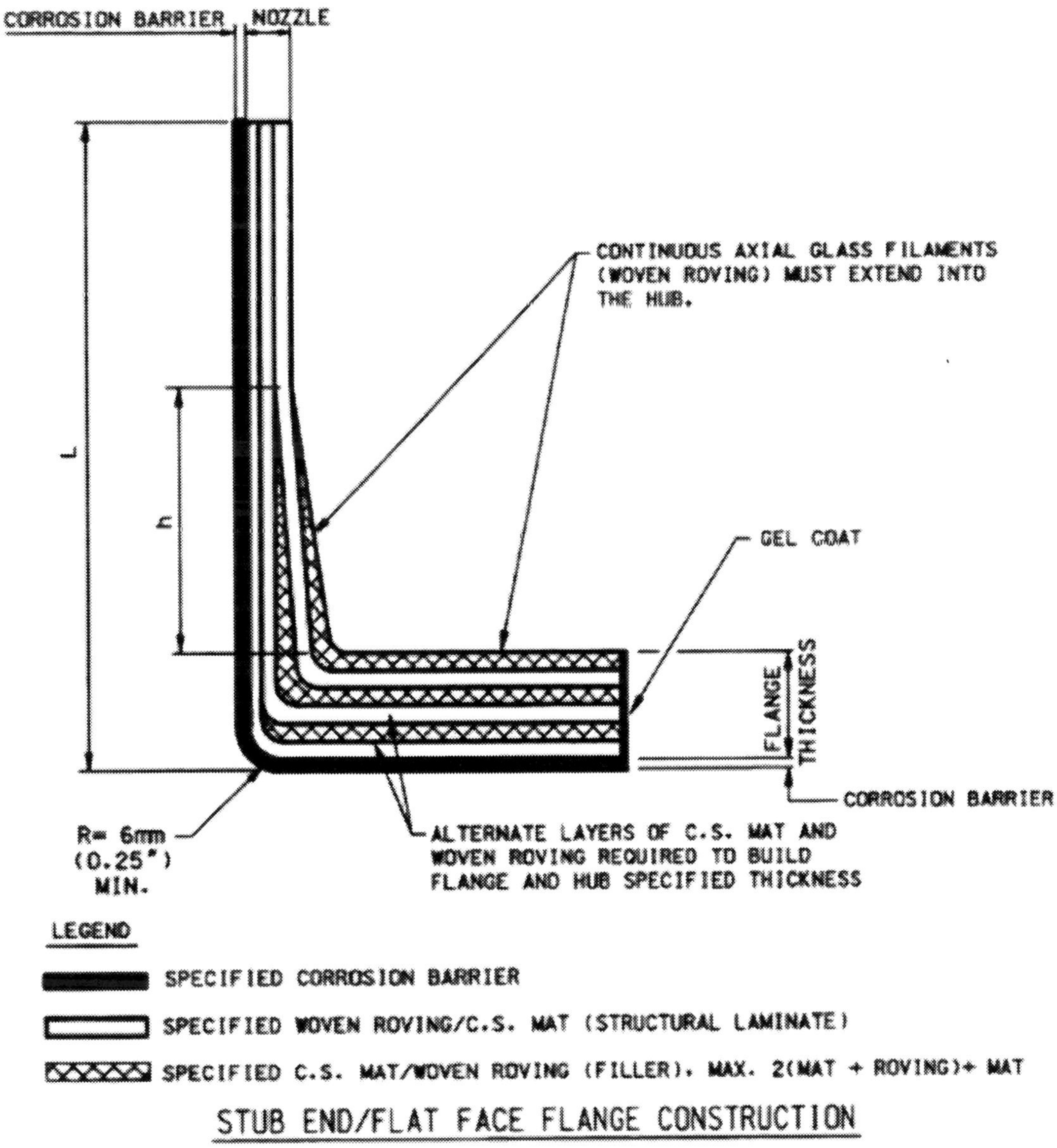

Figure 3-3. Stub End/Flat Face Flange Construction.

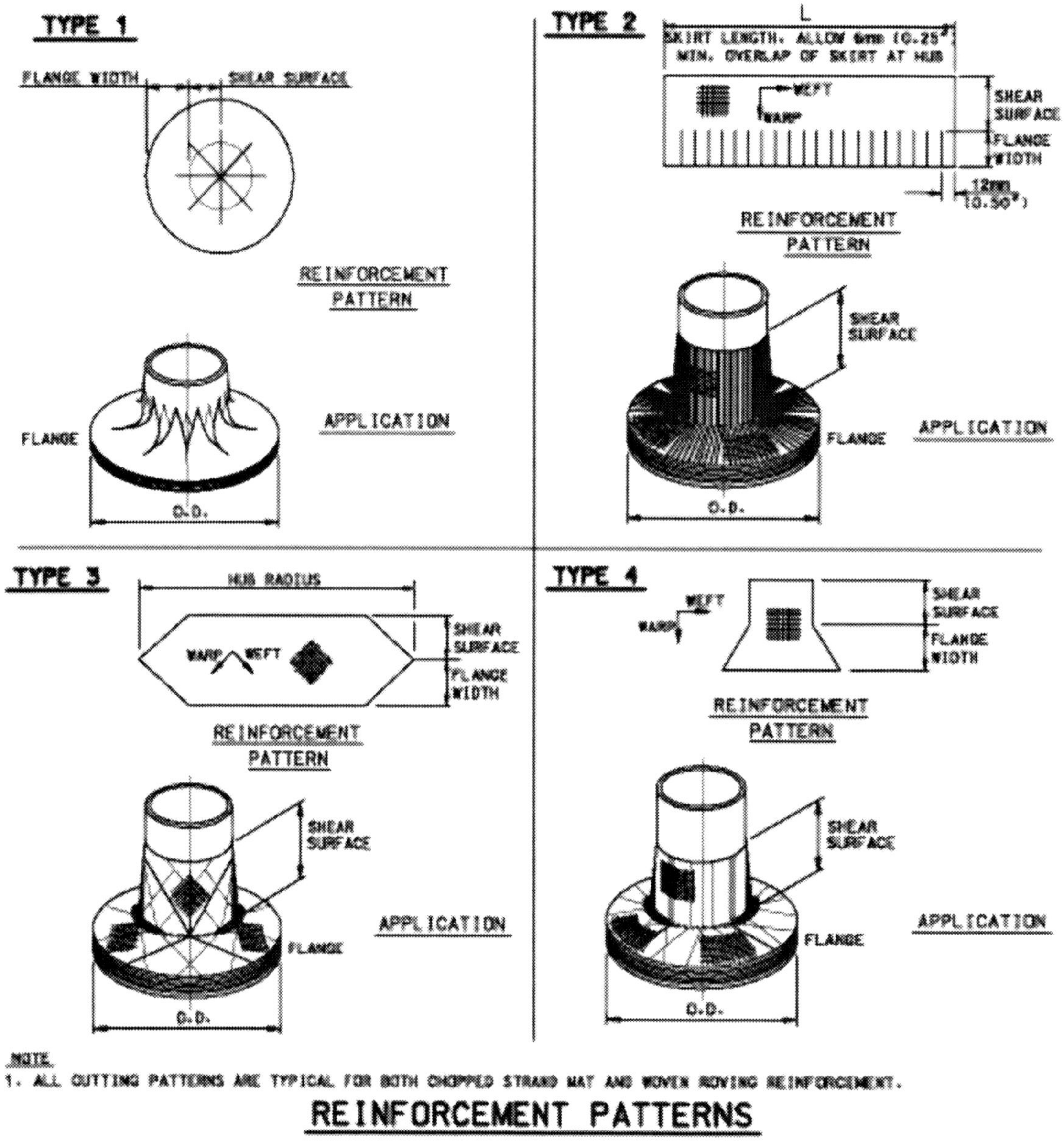

Figure 3-4. Reinforcement Patterns.

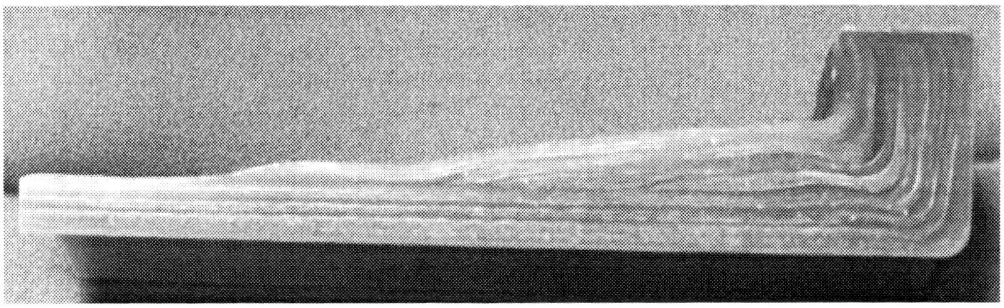

Figure 3-5. Cross Section of a Full-faced Flange Showing the Layers Continuing Up the Hub.

All of the standards discussed in Chapter 2 apply to contact molded flanges. In addition, ASTM D5421 is a standard specification for contact molded flanges made with glass fiber–reinforced thermoset resins. This standard provides requirements for materials, dimensions and tolerances, workmanship, and performance. The standard also describes design qualification test methods, including hydrostatic testing to four times the rated pressure of the flange and bolt torque testing to two times that recommended by the manufacturer. Flange manufacturers who have developed their own design methods as alternatives to the methods described in Chapter 2 will typically use ASTM D5421 to qualify their designs.

Advantages

- The equipment involved in contact molding is low-cost. The molds are inexpensive and easily modified, and the tools required may be nothing more than a bucket and roller for the resin and a drill for the bolt holes.
- There are several options for connecting a contact-molded flange to a pipe. Flanges may be molded directly onto an existing pipe (referred to as "two piece flange") or may be constructed integrally in a mold. In the latter procedure, the layers from the shell are continued into layers of a hub and flange during fabrication. Contact molded stub flanges may also be made as individual parts and connected to the host pipe as a bonded socket or with a wrapped butt-joint.

Disadvantages

- Contact molding is labor intensive and is not cost effective for the mass production of flanges.
- Resin concentration and dimensional uniformity, both within a part and from part to part, are difficult to maintain.
- Only addition-type crosslinking resins should be used, as the water from condensation-type resins (e.g., phenolics) is difficult to remove and can lead to voids.

Issues to Consider

Because the curing of a thermoset resin is an exothermic reaction, breaks should be taken in the lay-up process to allow heat to dissipate from the part. Applying too many layers without allowing the laminate to exotherm will cause delaminations between layers.[1] These delaminations, which can show as a cloudy region in the part, can degrade the mechanical properties of the flange if formed in the structural layer or compromise the corrosion resistance of the flange if formed in the corrosion barrier.

Another issue is flange face drawback. Because there is little or no pressure applied to a contact molded part during cure, the face of the flange has a tendency to pull back. As discussed in Chapter 2, excessive drawback may cause unanticipated bending stresses in the flange during bolt-up and could lead to cracking. NBS PS15-69, ASME B&PVC Section X, and ASTM D5421 specify limitations on the amount of flange face drawback. Drawback may be controlled by the choice of resin and hardening agent or by clamping the flange to a fixture that keeps the flange face planar during cure.[2] Another method of controlling drawback is to form an extra-thick corrosion barrier and then machine the barrier down after cure until the flange face is planar.[1]

Automated Techniques

Automated manufacturing techniques produce more consistent, higher quality flanges than contact molding. Because they are not as labor intensive, these methods are a better choice for high-volume production. However, the processing equipment can be costly, and the resin-rich corrosion barrier is still contact molded. In addition, flanges made from automated techniques are produced as individual parts and therefore must always be connected to the host pipe with a wrapped butt-joint, bonded socket joint, or the like.

ASTM D4024 is a standard specification covering "machine made" glass fiber-reinforced thermosetting resin flanges. The standard covers all flanges other than those made by contact molding. Similar to ASTM D5421, this standard provides requirements for materials, dimensions and tolerances, workmanship, performance, and qualification testing of FRP flanges made by using automated techniques. The design methods described in Chapter 2 should not be applied to flanges fabricated by using automated techniques. Flange manufacturers who have developed their own design methods can use ASTM D4024 to qualify their designs.

A description, along with some advantages and disadvantages, of several automated techniques is given below.

Filament Winding

Filament winding involves wrapping continuous strands of roving wetted with resin over a mold. Successive layers are wound over the mold until the desired thickness of the part is achieved. The part is then cured at room temperature or at elevated temperatures, depending on the process.

The filament winding process is typically used for making flange stubs for lap-joint flanges. The backup flange rings can be either metal or FRP. In the case of FRP backup rings, the layers must be oriented in such a way as to overcome the bending moments that the flange will experience during bolting (Figure 3-6) some manufacturers produce filament wound rings that are hoop wound. These rings are strong in the direction of the fibers but very weak in axial bending. Failures have been observed in those types of flanges either between the bolt holes or in the area between the bolt holes and the OD of the flange (Figure 3-7).

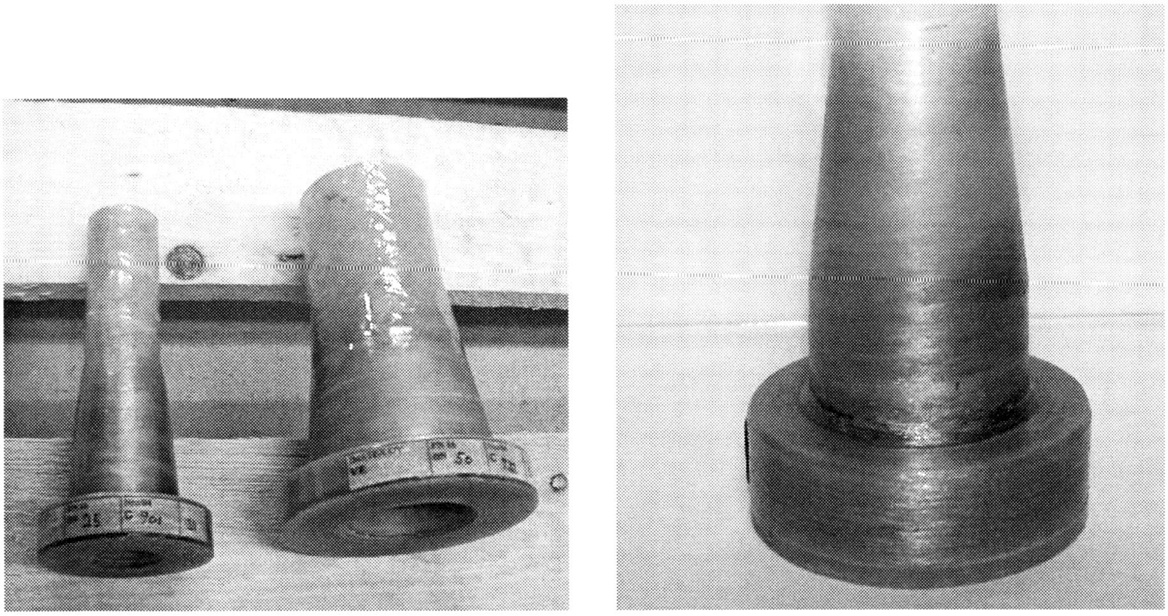

Source: Imatex, Finland

Figure 3-6. Filament Wound Lap-Joint Stubs.

Source: Albemarle Corp.

Figure 3-7. Filament Wound Backup Flange.

Because each layer of a filament wound flange has fibers oriented in the hoop direction, it is critical to understand the stresses that will develop in the flange during bolting and operation so that the flange will not be overstressed. Longitudinal bending stresses will occur at the flange/hub interface (Figure 3-8). Thus, to avoid cracking in the flange (Figure 3-9), the flange must either be designed carefully to ensure that there are enough helical layers to resist this bending stress, or layers of multidirectional mat and woven roving must be used for proper fabrication of a flange. This is one reason that only lap-joint stubs are manufactured with the filament winding process.

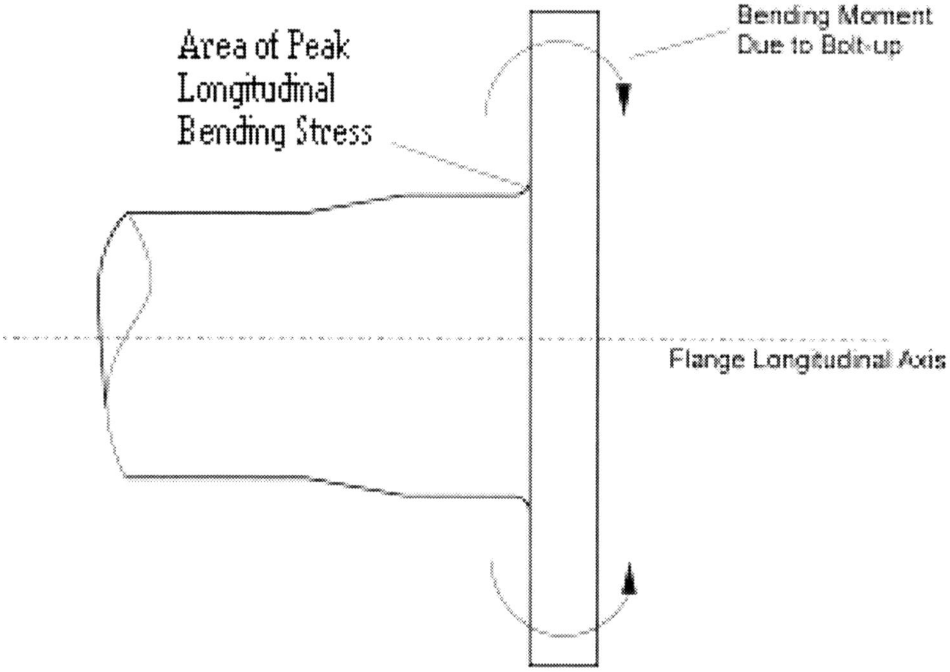

Figure 3-8. Flange Bending.

Source: The Dow Chemical Company

Figure 3-9. Delamination at the Base of the Flange.

Advantages

- Filament winding is a common method of fabricating FRP pipe, so manufacturers typically already have the necessary expertise to fabricate filament wound flanges.
- Large diameter flanges can be made relatively quickly.

Disadvantages

- For shear and bending stresses in the flange, it may be impossible to wind layers with optimal fiber orientation. Therefore, the number of layers must be chosen carefully to ensure that the final part has enough strength to resist all of the stresses that will be encountered in operation.

Compression Molding

In compression molding, a mix of fibers and resin is placed between male and female dies, which are pressed together to shape the compound while the resin cures (Figure 3-10). The fiber/resin mix is typically either bulk molding compound (BMC) or sheet molding compound (SMC). BMC, also referred to as premix, is a dough-like mixture of resin and chopped fibers. SMC is a sheet of slightly cured resin containing chopped or continuous fibers that is typically sold in rolls. It is also possible to fill the mold with preformed mat saturated with resin.[3] Compression molding machines are generally hydraulically driven and typically apply heat to cure the resin. The male and female dies are usually made of a hard metal, such as tool steel. Flanges are installed on the pipe by using a socket joint. This type of joint requires strict accordance to the specified installation procedure in order to be reliable. Many failures have been experienced in these types of joints because of poor installation practices.

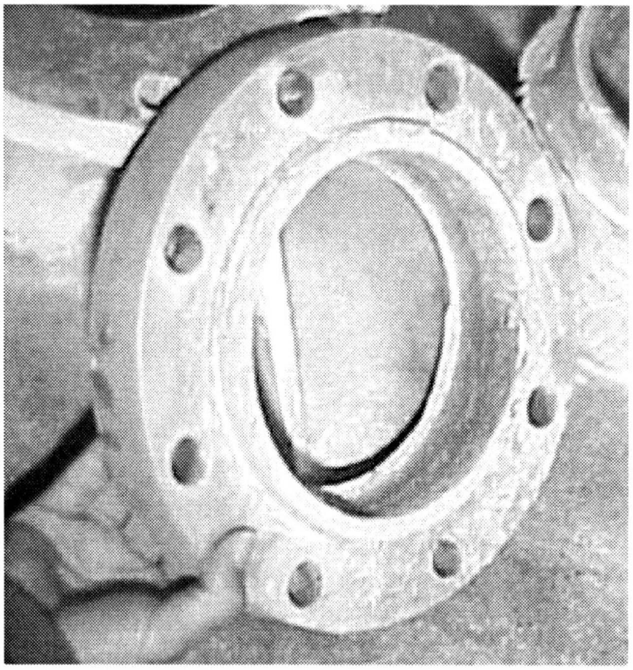

Source: Smith-Fibercast Corp.

Figure 3-10. A Compression Molded Flange.

Advantages

- Parts with tight dimensional tolerances can be made consistently.
- Bolt holes and custom features (grooves, etc.) can be molded right into the part.
- Pressure on the part during cure reduces flange face drawback.
- For SMC components, pressure on the part can also produce a higher density part (higher glass-to-resin ratio) and thus higher strength and stiffness.
- Production times can be very short.

Disadvantages

- Press equipment and dies are costly, usually limiting the process to small diameter flanges.
- Parts must be trimmed of excess material after the process.
- Flanges made with BMC are more susceptible to damage from over torquing.[4]

Resin Transfer Molding

Resin transfer molding (RTM) involves injecting resin under low pressure into a closed mold that contains the proper glass reinforcement material. The injection of the resin is typically accomplished with either a pump or a vacuum system. The mold design is critical to producing a quality part. The mold must be designed so that the resin flows easily to all areas with uniform concentration. In addition, the mold must have properly placed air vents to ensure that no air is trapped in the part as the resin is injected. Since the resin injection pressure is relatively low, many mold materials can be used in RTM as long as they hold their shape. Both plastic and metallic molds are often used (Figure 3-11).

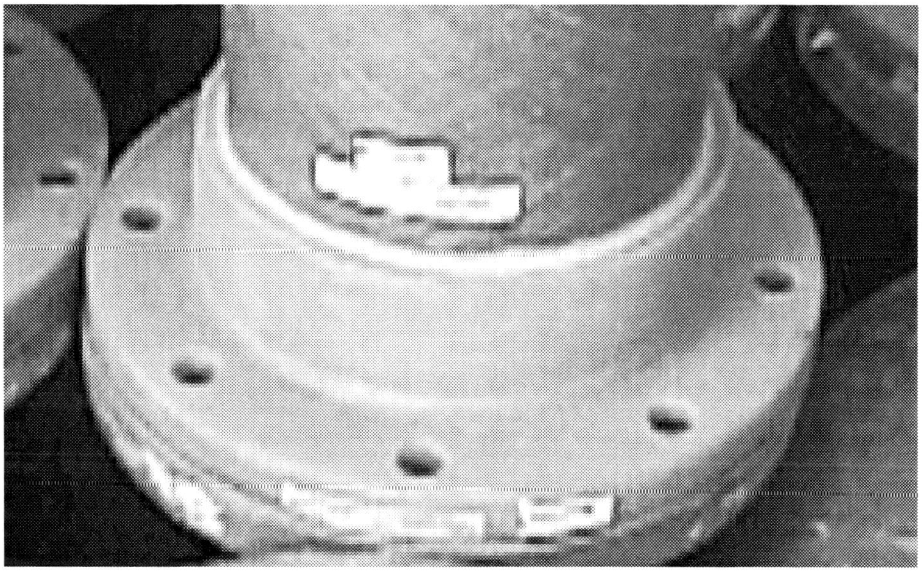

Source: Reinforced Plastic Systems, Canada

Figure 3-11.

Advantages

- Parts with tight dimensional tolerances can be made consistently.
- Bolt holes and custom features (grooves, etc.) can be molded right into the part.
- Part cures in a mold, reducing flange face drawback.
- Production times can be very short.

Disadvantages

- Molds are complex and must be designed carefully to ensure consistent resin concentration throughout the part.
- Molds are costly, usually limiting the process to small diameter flanges.
- Reinforcement movement in the mold during resin transfer can be a problem.

Technique	Advantages	Disadvantages
Contact Molding	• Equipment is low cost. • Flanges can be fabricated directly onto the host pipe, vessel, or tank.	• Process is labor intensive. • Resin concentration and dimensional uniformity are difficult to maintain. • Only addition-type cross-linking resins can be used. • Flange face drawback can be a problem.
Automated Techniques: Filament Winding Compression Molding Resin Transfer Molding	• Flanges are more uniform (both within part and from part to part) than contact molding. • Labor costs tend to be lower than contact molding.	• Equipment is expensive. • Standard design methods (see Chapter 2) are not applicable. • Flanges must be connected to the host pipe, vessel, or tank with a wrapped butt-joint or a bonded socket joint. • A resin-rich corrosion barrier typically cannot be produced on interior during the process.
Filament Winding	• Expertise is similar to that required to make FRP piping, vessels, and tanks. • Larger flanges can be produced relatively quickly and easily.	• Strength of the flange is limited to the direction of the fibers. Very weak in other directions. Careful design of flange is necessary to prevent cracking in the perpendicular to the fiber direction.
Compression Molding	• Produces flanges with tight dimensional tolerances. • Pressure on the part during cure reduces flange face drawback. • Production times can be very short. • Bolt holes and custom features can be molded into the part. • Flanges made using SMC process typically have higher strength and stiffness.	• Limited to small diameter flanges due to cost of equipment and dies. • Parts must be trimmed. • Flanges made with BMC are more susceptible to damage from over torquing. • Limited corrosion resistance because there is no corrosion resistant barrier.
Resin Transfer Molding	• Produces flanges with tight dimensional tolerances. • Part cures in mold, reducing flange face drawback. • Production times can be very short • Bolt holes and custom features can be molded into the part.	• Molds are complex and must be designed carefully. • Limited to small diameter flanges due to cost of equipment and dies. • Reinforcement movement in the mold during resin transfer can be a problem. • Limited corrosion resistance because there can be no corrosion resistant barrier.

Table 3-1. Comparison of FRP Flange Fabrication Techniques.

References

1. Private communication, Rick Watkins, Fibrex Corporation.

2. Blach, A.E. and Hoa, S.V., "The Effects of Pull-Back on Stresses in FRP Flanges," Exp. Tech., 1988, Vol. 12, pp.12-16.

3. Private communication, Robert Hawkins, Reinforced Plastic Systems.

4. Fiddes, D.E. and Lazarou, J., "Press Moulded Flanges for GRP Pipe Systems," Composites, Nov. 1972, pp. 254-257.

Bibliography

a. Strong, Brent A., Fundamentals of Composites Manufacturing: Materials, Methods, and Applications, © 1989 Society of Manufacturing Engineers.

b. Reinhart, T.J. et al., Engineered Materials Handbook Volume 1: Composites, © 1993 ASM International.

c. National Bureau of Standards, Voluntary Product Standard PS 15-69, Custom Contact-Molded Reinforced Polyester Chemical-Resistant Process Equipment, 1969.

d. ASME, Boiler and Pressure Vessel Code Section X, Fiber-Reinforced Plastic Pressure Vessels, 2003.

e. ASTM D4024-00, Standard Specification for Machine Made "Fiberglass" (Glass-Fiber-Reinforced Thermosetting Resin) Flanges, 2000.

f. ASTM D5421-00, Standard Specification for Contact Molded "Fiberglass" (Glass-Fiber-Reinforced Thermosetting Resin) Flanges, 2000.

4
Gaskets for FRP Flanges

One of the first steps in FRP flange design is to select the appropriate gasket. Proper gasket performance demands that the gasket be suitably compliant so as to conform to flange face irregularities and be sufficiently resilient to ensure that the flange-gasket contact stress always exceeds the internal pressure. The gasket material must also have an inherent stress-to-seal, which is reasonably lower than the actual developed gasket compressive stress. This chapter summarizes an approach to selecting an appropriate gasket for an FRP flange.

Gasket Factors

Various standards provide information about gaskets and gasket materials for FRP flanges. [1,2,3] For example, PS15-69 recommends that the gasket material have a durometer shore A or shore A2 hardness of 40 to 70. Also, the gasket thickness should be chosen as thin as possible to minimize the possibility of gasket blow-out by the internal pressure, but typically no less than 1/8 in (3 mm). Once the gasket has been selected, the gasket loading can be established. Initially, the gasket is pre-loaded by the bolt force and then the system is pressurized. When the internal pressure reaches the flange-gasket contact stress, separation of the flange and gasket occurs, causing leakage. For design purposes this complex loading history is broken into the seating and operating loading conditions. The seating condition considers the bolt pre-load, which provides the initial gasket pre-stress, while the operating condition includes the final internal pressure and corresponding bolt load.

Chemical resistance is one of the most important factors when selecting a gasket. Selecting the proper gasket also involves the following considerations:

- environmental concerns
- process safety
- expected service life of the flange
- maintenance costs

The chemical resistance of a gasket is often a strong function of temperature. A chemical resistance chart can be a helpful guide when selecting a gasket material, and most gasket manufacturers provide assistance with the gasket material selection. In some sensitive installations it might be necessary to perform field-testing to ensure that the chemical resistance and lifetime of the gasket are satisfactory.

There are two approaches for gasket selection: the *m* and *y* factors established by ASME,[2] and the newer design factors being developed by the Pressure Vessel Research Council (PVRC) for fugitive emissions. These approaches are discussed in the next two subsections.

M and Y factors

In the ASME B&PV Code,[5] the two gasket factors *m* and *y* are used in the analysis of flanged joints:

- The *m* factor defines how many times the residual load (original load minus the internal pressure) must exceed the internal pressure. In this calculation, the normal pressure and the test pressure should be taken into account.
- The *y* factor is defined as the minimum gasket stress required to sufficiently deform the gasket material into the flange face irregularities and ensure a tight, unbroken seal.
- It is important to note that these values were developed using much stiffer metallic flanges, thus imparting some degree of inaccuracy to their application for lower stiffness FRP flanges.

These factors depend on the material's sealing performance and are listed in the ASME B&PV Code[5] for different gasket materials. For elastomeric gasket compounds with Shore durometer readings below 75A, the *m* and *y* factors have the following suggested values: $m = 0.5$ and $y = 0.0$, and for elastomeric gasket compounds with shore durometer readings at or above 75A, the *m* and *y* factors have the following values: $m = 1.0$ and $y = 200$. These values are based on a combination of experience and testing. FRP flanges are typically designed with the minimum values of $m = 0.5$ and $y = 50$ for elastomeric materials and up to $m = 2$ and $y = 1200$ for some of the more compliant PTFE based materials.

Both the *m* and *y* factors vary with the type of gasket and the thickness of the gasket. Also, in many instances a good seal is obtained by the limited swelling caused by the reaction of the inside edge of the gasket with the fluid being contained. A certain amount of swelling can be desirable, as long as it reaches equilibrium and does not progress to a condition of degradation where the majority of the gasket starts to break down.

PVRC factors

The PVRC is currently developing new gasket factors (G_a, a, G_s) to replace *m* and *y*. These new factors incorporate the leak tightness of the joint as a design parameter.

This approach is still being developed and hence its use is not recommended until the standard is officially approved by the PVRC.

PT-Factor

Generally, the higher the temperature, the lower the allowable gasket working pressure. This is called the PT-factor of the gasket (pressure times temperature). Most gasket manufacturers publish the PT-factor for each of their gaskets. PT numbers are in the range of 20,000 to 50,000. These factors are seldom needed because the upper temperature limit of FRP materials do not pose a challenge; in other words, for a PT factor of 50,000 and a 150 psi (1 MPa) application, the maximum allowable temperature would be 330°F (165°C), and for a 100 psi (.69 MPa) application, the maximum allowable temperature would be 500°F (260°C).

Elastomeric Gaskets[10]

The type of elastomeric material that is most appropriate for a given FRP flange depends on the chemical compatibility of the gasket with the processing fluid. Also, gasket materials are sensitive to both temperature and pressure. As established in the preceding paragraph, the higher the temperature, the lower the allowable gasket working pressure.

When a gasket material is selected, it is recommended that an elastomer be used if an elastomer with sufficient chemical resistance characteristics can be found. The reason elastomeric gaskets are recommended is that they have better resilience and dimensional stability than other polymers, and require lower compressive stresses (*y* factor) to effectively seal.[4,6,7,8]

Some of the most common elastomers include[11]:

Natural Rubber or Polyisoprene (NR) - Offers good resistance to inorganic salts, ammonia, weak acids, and alkali; low resistance to oil and solvents. It ages due to ozone attack losing its strength and characteristics; not recommended for use in applications exposed to sun or oxygen. Outstanding resilience, superior resistance to tear and abrasion, very good tensile strength, excellent rebound elasticity (snap), good flexibility at low temperatures. Service temperature range: -60°F to 195°F (-50°C to 90°C).

Ethylene Propylene Diene Monomer (EPDM) - Excellent weatherability, good high and low temperature resistance, low gas permeability. Resistant to ozone, steam, water, alcohols, brake fluids. Poor resistance to petroleum oils, diester lubricants. Temperature range: -60°F to 280°F (-50°C to 138 °C).

Chlorinated Polyisoprene Rubber (CR or Neoprene) - Good general purpose elastomer, with moderate oil, ozone and water resistance. Resistant to freons, high analine points oils, mild acids. Not recommended for phosphate esters, ketones, brake fluids. Temperature range: -65°F to 260°F (-54°C to 127 °C).

Nitrile (NBR) - Excellent resistance to petroleum products, excellent compression set; resistant to oil, water, silicone greases, and oils. Not recommended for halogenated hydrocarbons, ketones, ozone, and sunlight. Temperature Range: -65°F to 275°F (-54°C to 135 °C).

Styrene Butadiene Rubber (SBR) - General purpose elastomer, good adhesion to various substrates, good abrasion resistance, good resiliance. Resistant to brake fluids, alcohols, water. Not recommended for solvents, oil/gas, sunlight. Temperature range: -65°F to 225°F (-54°C to 107 °C).

Silicone (SI) - Best overall temperature range of elastomers currently available. Excellent resistance to dry heat, as well as resistance to low temperature, brake fluids, high analine point oils. Not recommended for some dynamic applications due to poor tensile strength and tear. Poor in petroleum oils and ketones. Temperature range: -150°F to 500°F. (-101°C to 260 °C).

Fluoro Elastomer (FPM or FKM, also known as Viton®)- Outstanding resistance to the effects of heat, chemicals, and solvents. Viton® has good tensile resistance and low compression set. Limitations include poor low-temperature properties, limited hot-water resistance and shrinkage. Temperature range: -50° F to 600°F (-45°C to 315 °C).

Recommended thickness of gaskets for most FRP flanges is 1/8 in (3 mm); however, larger diameter flanges may warrant thicknesses up to 1/4 in (6 mm), depending on the application.

® Registered Trademark of E.I. Dupont

Non-Elastomeric Gaskets

The most prevalent types of non-elastomeric gaskets utilize polytetrafluoroethylene (PTFE) or bi-directionally expanded PTFE gaskets. These gaskets typically have less elasticity and resilience than elastomers, but they have the advantage of being stable in virtually any chemical environment. They also have higher mechanical strength and blowout resistance, and in some cases even have higher compressibility than do elastomeric materials. The main drawback of PTFE-based gaskets is the higher required stress to seal (y factor), translating into higher flange stresses. The recommended design approach for these gaskets is to use the PTFE gasket manufacturer's m and y values to determine the required loads.[4, 6, 7, 8]

Bi-directionally expanded PTFE is a type of PTFE that has the chemical resistance benefits of conventional PTFE but with much less creep and cold flow. It conforms to rough sealing surfaces, yet compresses into a tough gasket that makes a long lasting seal for all types of flanges and fluids.

Installation procedures for PTFE-based gaskets are available from the gasket manufacturer. Typical recommendations include:

- Lubricate fasteners with an anti-friction compound, such as molybdenum disulfide or a copper/nickel anti-seize compound, to produce more uniform conversion of torque into clamping force (less statistical scatter of developed clamping force). Lubricant should be applied to the fastener threads and all bearing surfaces (underside of bolt heads, nuts, and washers). Lubricant should never be applied to the gasket surfaces.
- Use a torque wrench to tighten fasteners, in order to achieve a more uniform clamping force on each fastener.
- Use a "star" or "criss-cross" stepwise torque pattern to ensure even compression of the gasket and avoid low compressive stress zones due to cocking of the flanges. Two torque increments commonly used by industry and recommended by gasket manufacturers are 20% -> 50% -> 100%, and 30% -> 60% -> 100%, with the first increment sequence often used for softer gaskets.
- Bolts should be torqued to the highest level specified by the flange manufacturer. In many cases, flange manufacturers will allow higher bolt torques than those values published because these are typically based on the use of an elastomeric gasket, and not on flange stresses.
- If the piping system is operating above 150°F (65°C), or if PTFE-based gaskets are used, the bolts should be re-torqued in a star pattern sequence (see Chapter 5) after 24 hours or one thermal cycle.

Other Gasket Designs

A variety of gaskets have been designed to seal at the low seating stress achieved by non-metallic flange applications, e.g., FRP and carbon. These include metallic or plastic-reinforced elastomer gaskets where an encapsulated metal or plastic core is used to provide additional stiffness, memory, or blowout resistance. The reinforcement often consists of a corrugated or perforated metal foil core. Similarly reinforced PTFE-based gaskets are also available. Alternatively, there are several manufacturers which offer a fluoropolymer-encapsulated elastomer gasket design which provides a better chemical resistance, while retaining most of the compliance and low stress-to-seal of an elastomer. Encapsulated or plain elastomer gaskets may include multiple circular sealing ribs (Figure 4-1) on their sealing surfaces (as described previously in Chapter 2 – FRP Flange Design) to provide increased or peaked localized compression to further improve the sealing characteristics. One word of caution is that gaskets with sealing ribs should not be used with FRP flanges that also have sealing ribs, as imperfect centering of the gasket could result in overlap of the mating circumferential ribs and compromise the seal. Finally, reduced-area PTFE-based gaskets are also available (Figure 4-2) and make use of reduced thickness areas (circular rings) to substantially increase the developed compressive stress on the gasket material, similar in concept to the sealing rib approach but with a more pronounced increase in stress.

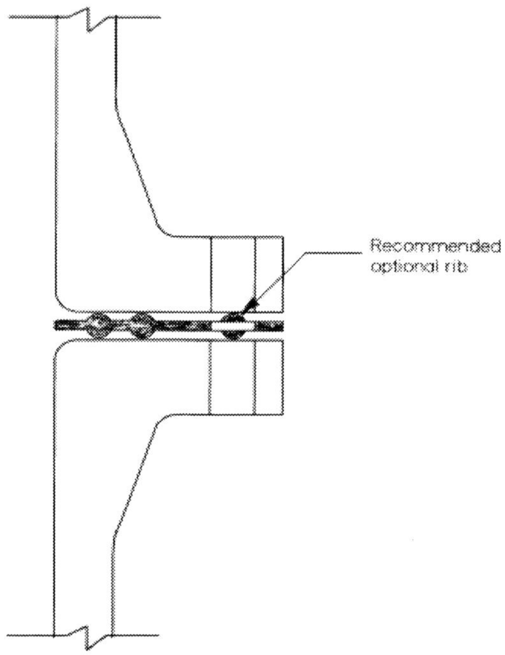

Figure 4-1. Full-Faced Gasket with Sealing Rings.

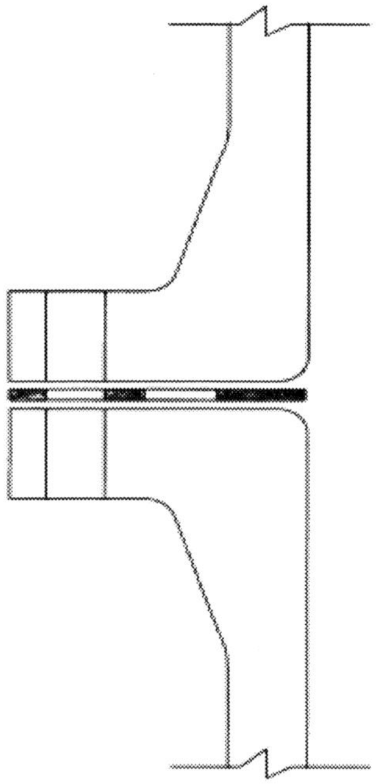

Figure 4-2. Full-Faced Gasket with Reduced Contact Area.

"O" Rings

"O" ring gaskets, although not very common in FRP flanges, require special design. There are applications that require high pressure large diameter FRP flanges that can be best handled by using "O" ring gaskets. These flanges require special molding and tooling and are typically fabricated by the filament winding process. Some fabricators[12] recommend using "O" ring gaskets for all flange diameters larger than 42 in (1066 mm). (Figure 4-3).

Courtesy of: Gulf Eternit (Future Pipe Industries)

Figure 4-3. Large Diameter Filament Wound Flange with "O" Ring Gasket.

Discussion

All gaskets will relax with time, causing a loss of clamping force, and thus a corresponding decay in compressive stress acting on the gasket. This phenomenon is caused by viscoelastic and viscoplastic creep of the polymer gasket material and will typically be higher for PTFE-based gaskets than for elastomeric gaskets due in large part to the cross-linking that is present in the elastomers and absent in fluoroplastics.

If the loss of clamping force is sufficient, then the compressive force holding the gasket in place can fall below the internal pressure and a leak or blow-out can occur. Stated a different way, creep relaxation degrades the compressive stress acting on the gasket over time. Should the compressive stress required for a particular application drop below the leak threshold, a leak will likely occur. From a root cause standpoint, there are a number of ways to either reduce the creep relaxation that will occur or offset the expected relaxation:

- Retorque the bolts after one thermal cycle or 24 hours (offset relaxation)
- Use conical spring (Belleville) washers (offset relaxation)
- Increase the bolt load if the flange design permits (offset relaxation)
- Use circular sealing rib design on either the flange or gasket (offset relaxation)

- Use a thinner gasket to reduce relaxation (may not be practical with FRP flanges because of the tolerances in fabrication)
- Use a denser gasket material (reduce relaxation)
- Use a gasket material with better creep relaxation properties (reduce relaxation)

There are many different reasons for gasket failures. The most common are as follows:

- Uneven loading of flanges due to:
    - Flange face misalignment (non-parallel)
    - Improper torquing or sequence (not following industry standard flange make-up procedures, e.g., torque sequence, incremental torque, use of torque wrench, not allowing joint relaxation time.)
    - Flange warpage
    - Pipe strain
- Gasket compressive load too low, due to:
    - Inadequate initial torque value
    - Lack of fastener lubricant
    - Creep relaxation of gasket material
- Vibration loosening of fasteners
- Operating piping loads
- Thermal cycling or thermal excursions
- Operating at cold temperatures
- Bolt strength too low, resulting in yielding of bolts during torquing
- Improper gasket installation or gasket damage
- Pre-existing flange sealing surface damage
- Chemical or thermal degradation of the gasket material
- Bolt weakening due to corrosion

In summary, to reduce gasket failures it is important to:

- Properly select and install the gasket following the gasket manufacturer's instructions
- Bring the flanges together slowly and parallel
- Lubricate the fasteners and bearing surfaces
- Torque flanges using proper torque values and criss-cross stepwise patterns
- Retorque flanges after one thermal cycle or 24 hours where warranted

- For applications where creep relaxation is especially problematic, use conical spring washers on each bolt to offset the relaxation

References

1. National Bureau of Standards, Voluntary Product Standard PS 15-69, Custom Contact-Molded Reinforced Polyester Chemical-Resistant Process Equipment, 1969.

2. ASME, Boiler and Pressure Vessel Code Section X, Fiber-Reinforced Plastic Pressure Vessels, 2003.

3. British Standards Institution, BS 4994-87, British Standard Specification for Design and Construction of Vessels and Tanks in Reinforced Plastics, 1987.

4. Product literature and private communication, W. L. Gore & Associates, Inc.

5. ASME, Boiler and Pressure Vessel Code Section VIII, Appendix 2, 2003.

6. Product literature and private communication, Garlock Sealing Technologies.

7. Product literature and private communication, C.T.G. Inc.

8. Product literature and private communication, Gasket Resources Inc.

9. Product literature and private communication, W. L. Gore & Associates, Inc.

10. Industrial Gaskets, Jose Carlos Veiga, Teadit N.A.

11. Thomas A. Caserta, Inc.

12. Gulf Eternit (FPI Group), Dubai UAE.

5

FRP Flange Bolt-up

The final stage of creating a flanged joint is bolt-up. Preloading the bolts properly is critical as the bolts must exert enough clamping force on the gasket to prevent leakage, yet not so much force that the flange or bolts themselves are damaged. This chapter will summarize common bolt materials and lubricants used with FRP flanges, methods of determining the amount of torque required to achieve the proper bolt preload, and some strategies for bolting up dissimilar flange faces.

Bolt Materials and Lubrication

The end user often chooses the bolt material for flanged joints. However, if the flange manufacturer used one of the design methods discussed in Chapter 2, then a bolt material, with known ultimate tensile stress, may have already been assumed in the flange design. Therefore, if the bolt material is not already specified, the end user should consult the flange manufacturer before bolting up the flange.

Corrosion-Resistant Bolt Materials

FRP piping, vessel, and tank systems are often used in corrosive environments. Hence, bolt materials chosen for FRP flanged joints are often corrosion resistant. Table 5-1 lists some common corrosion-resistant bolt materials along with a corresponding ultimate tensile strength range.

When a bolt material is chosen, its ultimate tensile strength or yield strength should be considered. In particular, for large diameter full-faced flanges, the bolt loads required for gasket seating and operating conditions may be very high and a material that can handle these loads must be chosen. If one of the methods described in Chapter 2 is used to design the flange, the allowable bolt stress, which is a required parameter, can be based on the ultimate tensile strength with a generous safety factor.

Washers must always be used under nuts and bolts in full-faced flanges to distribute the compressive forces that can crush the FRP part.[1, 2] ANSI B18.22.1, Type A narrow flat washers are specified in ASME Boiler and Pressure Vessel Code Section X, but FRP flange manufacturers sometimes provide their own specification.

Lubricants

All FRP manufacturers recommend that the bolts be well lubricated, along with all bearing surfaces (under bolt heads and nut and washer faces). Reducing the friction between the bolting components increases the amount of torque that is converted to actual clamping force. As much as 90% of the torque applied to a fastener is consumed by bolt or nut face and thread friction, leaving only 10% or so to generate tension in the fastener and thus a clamping force.[5] Lubricant should never be applied to the gasket surfaces. Lubricants also help prevent galling between the nut and the bolt at bolt-up.

ASTM D4024 and D5421 recommend that a solid lubricant be used, which is appropriate for bolting up FRP flanges as high bolt loading and relatively low torque are involved.[3] Common solid lubricants used in bolting FRP flanges are discussed below.

- Molybdenum disulfide (MoS_2) – A dark gray crystalline material that has been used as lubricant for decades. It has excellent adhesion to metals and good load bearing capacity. It can be applied as a loose powder or as a bonded coating. In the latter case, the MoS_2 powder is incorporated into an adhesive such as a polymer, natural resin, or molten solid.
- Polytetrafluoroethylene (PTFE) – A white solid polymer that lacks chemical reactivity and is nontoxic. Its load-carrying capability and endurance limit are relatively low compared to those of MoS_2. It can be used as a bonded coating, but its low friction and chemical inertness make it difficult to bond to bolt materials. It is rarely used as a powder.
- Other polymers – Fluoropolymers such as polyvinylflouride (PVF_2) and polytrifluorochloride (PTFCE) can bond to bolt materials more easily than PTFE, resulting in stronger bonded coatings. However, this property comes at the expense of low friction.
- Nickel- and copper-based lubricant and anti-seize metallic pastes are also commonly used for fastener lubrication. These pastes will often produce a nut factor (K) of 0.13 to 0.16. Additional discussion on nut factors and tabular data can be found later in this chapter.

Material	Examples	UTS	Notes
Carbon Steel	A193 (except B series) A325, A490, SAE J429, etc.	80-125 ksi	Can be made somewhat corrosion resistant by painting or coating with a thin film of fluoropolymer material.
Austenitic Stainless Steel	A193 B8 series, A320 B series, 303, 304, 316, etc.	75-120 ksi	Most common of the stainless steels and has more corrosion resistance than the others. Can't be heat treated but can be cold worked. Susceptible to chloride stress corrosion cracking and galling.
Ferritic Stainless Steel	430, 430F	70 ksi	Can't be heat treated or cold worked.
Martensitic Stainless Steel	410, 416, 431	70-180 ksi	Heat-treatable. Can experience stress corrosion if not properly treated.
Nickel-copper	Monel 400	70-80 ksi	Can be cold worked but not heat treated.
Nickel-copper-aluminum	Monel K500	130 ksi	Can be both cold worked and heat treated.
Titanium	Ti6Al-4V	135-200 ksi	Expensive. Aerospace alloy.
Superalloys	MP35N, Inconel® Alloys	145-286 ksi	Some are virtually immune to marine environments and stress corrosion cracking. Expensive.
Nylon		11 ksi	Outstanding corrosion resistance but low strength.

Source: Bickford, John H. An Introduction to the Design and Behavior of Bolted Joints, 3rd ed. pp 128-129.

Table 5-1. Corrosion-Resistant Bolt Materials.

Bolt Torque

Applying the proper amount of torque to bolts, in the proper sequence, is critical to achieving the required and uniform bolt preload and, hence, is essential to ensuring the sealing integrity of an FRP flange connection. The torque required can be determined through the flange manufacturer's recommendation or can be calculated, provided enough information is known about the flange design parameters. There are two important aspects related to a gasketed joint. The first is the compressive stress (and corresponding bolt load) required to adequately seat a given gasket in a given application.

The second is the allowable bolt load that an FRP flange can safely carry without imparting permanent damage. To state the obvious, it is very important that the required bolt load for seating the gasket be less than the flange rating. If not, a lower stress-to-seal gasket or a stronger FRP flange will be required. This is not normally a problem when very soft elastomeric gaskets are used, but it can be an issue for harder elastomeric or PTFE-based gaskets.

Torque Tables

Many FRP flange manufacturers publish recommended torque values for their flanges in the form of torque tables. These tables typically list the recommended bolt torque for a given flange type, size, and pressure rating and usually assume that a specific gasket type and hardness will be used. An example is given in Figure 5-1, which lists the maximum bolt torque for full-faced FRP flanges produced by Bittner Industries, Inc.[4] Note that this table is for illustrative purposes only and is not intended as a general guide. Flange manufacturers should always be consulted for specific torque recommendations appropriate for their flanges.

MAXIMUM BOLT TORQUE IN FT. LBS.

Pipe Dia.	INTERNAL PRESSURE RATING PSIG					
IN.	25	50	75	100	125	150
2	25	25	25	25	25	25
3	25	25	25	25	25	25
4	25	25	25	25	25	25
5	25	25	25	25	25	30
6	25	25	25	25	35	40
8	25	25	30	40	50	60
10	25	25	30	40	50	70
12	25	25	35	45	60	80
14	25	30	40	60	75	100
16	25	30	50	70	80	
18	30	35	50	80	100	
20	30	35	60	90		
24	35	40	70	90		
30	40	45	80			
42	50	55				
48	60					

The above figures are based on full-faced gaskets having a Shore A Durometer 60-70; 1/8 in gasket thickness for 24 in diameter and under 3/16 in in gasket thickness for 24 in diameter to 42 in diameter. and 1/4 in gasket thickness 48 in diameter and over.

Figure 5-1. Bittner Industries, Inc. Bolt Torque Recommendations (for Illustration Only).

Torque Calculations

The maximum and minimum torque can be calculated if the required bolt loads for gasket seating and under operating conditions, as well as several design parameters of the FRP flange, are known. ASME RTP-1, Shigley[3], and Bickford[5] describe formulae for torque calculation, as summarized below.

ASME RTP-1

ASME RTP-1 presents formulae for the minimum and maximum bolt torque in Appendix NM-12. The minimum bolt torque is given by

$$T_{min} = \frac{\sigma_{bm} A_b D_b}{60} \quad \text{(ft-lbs)}$$

which is a simplified version of

$$T_{min} = \frac{K_{bm} A_b D_b}{12} \quad \text{(ft-lbs)}$$

and assumes that the nut factor K = 0.2, which is the default assumed value for most applications. If a different nut factor is warranted for a specific application, the standard formula constants used in the subsequent equations should be adjusted accordingly. In addition, σ_{bm} is the minimum bolt stress and is the larger of the bolt stresses required to seat the gasket σ_{bg} and the bolt stress under operating conditions σ_{bo}, which are calculated by using the gasket *m* and *y* factors during the flange design procedure (see Appendix B for these stress formulae). A_b is the bolt root area and D_b is the bolt diameter. These equations assume inches for dimensions and psi for stress.

The maximum bolt torque must be lower than the torque that will result in either the FRP material or bolt reaching their design stress limits. The torque that results in the FRP material design stress limit is given by

$$T_F = \frac{\sigma_{bf} A_b D_b}{60} \quad \text{(ft-lbs)}$$

Where σ_{bf} is the bolt stress that causes the flange to reach its design stress (see Appendix B for this stress formula). The torque that will result in the bolt design stress limit is given by

$$T_b = \frac{\sigma_b A_b D_b}{60} \quad \text{(ft-lbs)}$$

where σ_b is the bolt design tensile stress. The maximum bolt torque is the smaller of T_f and T_b. In practice, the torque applied to the bolts should be closer to this calculated maximum than to the calculated minimum.

Appendix B contains torque calculations using the method of ASME RTP-1 for the example flange design discussed in Chapter 2.

Simplified Approach[5]

A simple formula for calculating the torque required to achieve a bolt preload is

$$T = F_P KD_b \ \text{(ft-lbs)}$$

where F_P is the target bolt preload, K is the nut factor, and D_b is the bolt diameter. Note that this nut factor (coefficient of friction) K depends on surface finish, coatings, and thread lubrication. Typical values for K are:[8]

- 0.2 for untreated fasteners
- 0.15 for coated fasteners
- 0.06 for molybdenum disulfide or Teflon® lubrication

Using this formula, the minimum and maximum torque can be determined if the minimum and maximum target bolt preloads are known. For the case of an FRP flange, the minimum target bolt preload can be found by multiplying the aforementioned minimum bolt stress σ_{bm} by the bolt root area A_b (see Table 5-2). The maximum target bolt preload can be taken as the lower of the bolt loads that will result in either the FRP material or the bolt reaching its design stress limits. These bolt loads can be found by multiplying σ_{bf} and σ_b, respectively, by the bolt root area A_b.

The tightness factor takes into account such factors as the tool accuracy, operator skill, lubricity, condition of the threads, etc. An accurate measurement of nut factor requires laboratory experimentation with the exact flange joint geometry and conditions seen in the field. However, experimental nut factors have been published and can be used as order-of-magnitude indications for given lubricants. See Table 5-2 for typical values of K.

Nominal Diameter of Bolt	Number of Threads per Inch	Tightening Factor K	Cross Sectional Area of Thread (in^2)
1/4	20	0.23	0.027
5/16	18	0.22	0.045
3/8	16	0.18	0.068
7/16	14	0.19	0.093
1/2	13	0.2	0.126
9/16	12	0.21	0.162
5/8	11	0.19	0.202
3/4	10	17	0.302
7/8	9	0.17	0.419
1	8	0.18	0.551
1 1/8	7	0.2	0.693
1 1/4	7	0.19	0.890
1 3/8	6	0.2	1.054
1 1/2	6	0.18	1.294
1 5/8	5	0.19	1.515
1 3/4	5	0.2	1.744
1 7/8	5	0.21	2.049
2	4	0.19	2.3

Table 5-2. Steel or Alloy Bolts or Studs.

Torque Sequence

The bolt torque sequence should follow that prescribed in ASTM D4024 and D5421 and shown in Figure 5-2. This is called the "star" or "criss-cross" pattern and, when followed, provides a better chance of achieving a uniform load for all of the bolts.

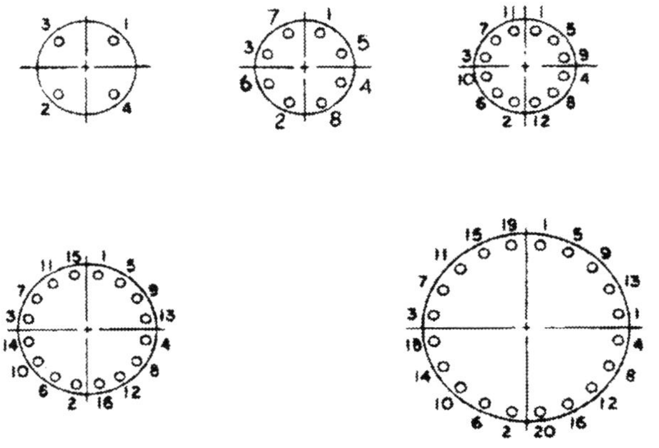

FIG. 2 Bolt Torquing Sequence

Figure 5-2. Bolt Torque Sequence.

The bolts should be "snugged up" by hand, perhaps to 10% of the final torque as a guideline, prior to starting the torquing process. This step is especially important when soft gaskets are used in order to prevent cocking the flanges. Once complete, a "criss-cross" or "star" pattern should be used in a stepwise fashion to tighten the bolts. Some flange manufacturers and most gasket manufacturers specify the incremental amount of torque that should be applied to each bolt in the "star pattern" during bolt-up. For example, in their latest Integraline design manual, Fibrex specifies that bolts should be initially torqued in the "star pattern" to 50% of the recommended value for their flanges and subsequent tightening should be in the same "star pattern" until the recommended torque level is achieved. In their installation manual, Reinforced Plastic Systems lists specific torque increments by flange size for both their full-face and lap-joint flanges. Two torque increments commonly used by industry and recommended by gasket manufacturers are 20% → 50% → 100% and 30% → 60% → 100%, with the first increment sequence often used for softer gaskets.

Issues to Consider

The accuracy of the preload placed on a bolt through the torquing process depends on many factors, including the lubricity, the condition of the threads, and the type of tightening tool used. For example, the preload on a bolt may vary by up to 30% for a given torque generated by using a manual torque wrench.[5]

When bolts are tightened, there is elastic interaction between the bolts. That is, after tightening any bolt, tightening another bolt will decrease the bolt load in the previously tightened bolts. Even under controlled laboratory conditions, it is very difficult to achieve uniform bolt loads for all of the bolts in a flanged joint. Uniform bolt load is necessary to properly control gasket stress and ensure the integrity of the flanged joint. Following the "star pattern" and applying torque to the bolts in stepwise increments (up to ten[6]) will help alleviate the problem of bolt load non-uniformity. Materials Technology Institute (MTI) has found that two additional circular torque passes following the full torque "star" pass, either clockwise or counterclockwise, will produce especially uniform clamping forces from bolt to bolt.[8]

Bolt-up of Dissimilar Flange Faces

FRP material is not as ductile nor as stiff as most metals and will crack when forced to undergo relatively large deformations. Due to this characteristic, it is imperative to ensure that an FRP flange face is fully supported across its diameter when bolting an FRP flange to a flange with a dissimilar flange face. For example, if one were to attempt to bolt a full-faced FRP flange to a raised-face metal flange, the unsupported regions of the FRP flange face would be subject to bending deformations and would be susceptible to cracking (see Figure 5-3).

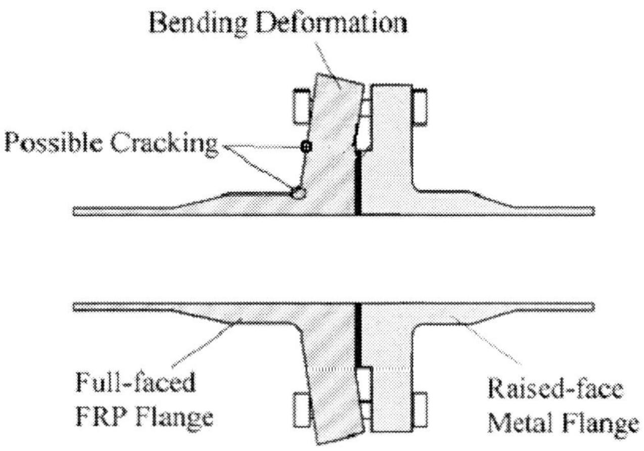

Figure 5-3. Bending Deformations When Bolting Dissimilar Flange Faces.

To avoid these large bending deformations, a spacer should be placed between the flange faces to fully support the FRP material. For example, the Dow Chemical Company uses a metallic spacer lined with a polymer such as polypropylene, Tefzel, FEP, or PTFE to ensure corrosion resistance as shown in Figure 5-4.[7] Another strategy that Dow has developed is a rubber "filler ring" (such as EDPM) that is placed in the outer diameter of the raised face flange (see Figure 5-5). This eliminates the use of a metal spacer, which can corrode. The "filler ring" should be equal in thickness to the raised face so that it stays flush with the raised face under the compressive forces of the bolts. A third solution (if space is available) to the problem is accomplished by building a short spool that has a full-faced flange on one end and a lap joint on the other (see Figure 5-6).

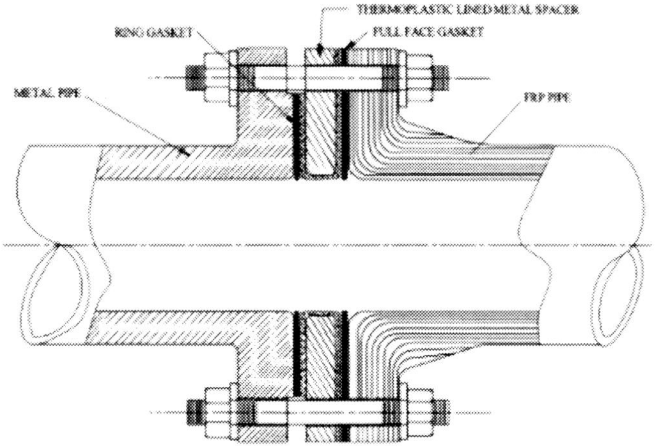

Figure 5-4. Dissimilar Flange Faces with Metallic Spacer.

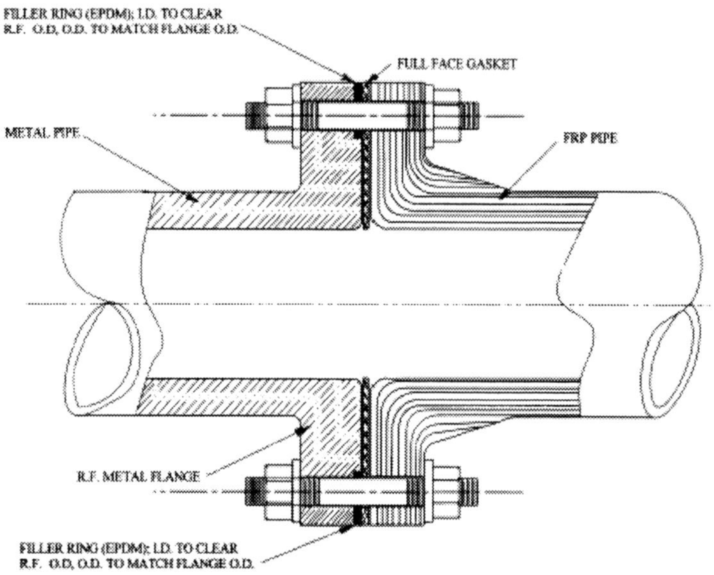

Figure 5-5. Dissimilar Flange Faces with Filler Ring.

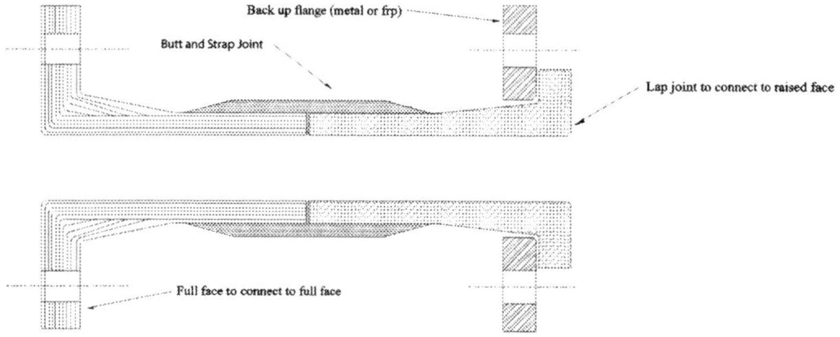

Figure 5-6. FRP Transition Spool (FF to RF).

Flange Disassembly Issues

At some point during the service life of an FRP flange pair, it will be necessary to disassemble the flange, regardless of whether it is for maintenance of other components in the piping system, to replace a degraded gasket, or for internal inspection of the piping. While in general the disassembly procedures are similar to those for metallic flange joints, there are several considerations unique to FRP. General guidelines containing these FRP-specific considerations are provided below for the disassembly of FRP flanges:

Bolt/Fastener Removal

It is not uncommon for bolting used in an industrial outdoor environment to suffer corrosion, particularly since FRP piping is often used in and around corrosive processes. In the more severe environments (marine, wet, splash and spill areas, etc.), corrosion can result in failure of the fasteners if the material of construction is not sufficient and/or if the fasteners are not inspected at appropriate intervals. In any case, corrosion of the exposed fastener threads will make disassembly that much more difficult. Guidelines for disassembly are as follows:

- It is best to "break" the torque on all bolts (loosen) before backing the nuts off of any one bolt.
- Use a back-up wrench to prevent gouging of the flange under the washers or in the bolt hole.
- Do not use heat to disassemble bolts as heat can damage the FRP.
- Use a thread penetrant like Liquid Wrench™ as needed for rusted fasteners.
- When all else fails, either use a nut splitter or a cold cutting technique to separate the fasteners.

As a general guideline, bolt and stud length should preferably provide a projection beyond the back face of the nut of no less than flush and nominally 2 threads. Due to flange thickness variations, this may not always be achievable.

The amount of thread surface exposed and prone to corrosion should be minimized. For highly corrosion-resistant fasteners, this concern is essentially voided. If fastener corrosion is a problem, consideration should be given to the use of a corrosion-resistant coating/plating or upgrading the material of construction of the fasteners. See earlier sections of this chapter for more discussion on this subject.

Gasket Removal

After the bolts have been removed, the flanges can be separated and the gasket removed. Under no circumstances should any type of pry bar [the use of a plastic (e.g., PP, PE) spatula type may be useful…] be used to separate the flanges, as the point loads will most surely damage the sealing surfaces, particularly if the flanges have sealing ribs. Plastic spatula-type pry bars may be an acceptable exception. Solvents should be avoided for cleaning of the flange faces, unless they are used sparingly or the FRP resin compatibility is confirmed in advance. During cleaning it is important to avoid damaging the sealing ribs on flanges that have this feature. These sealing ribs are typically more fragile than the surrounding surfaces, and if sufficiently damaged, can actually compromise sealability.

Flange Inspection

Once the flanges are clean, and prior to reassembling any flange pair, it is important to inspect the flanges for damage, which might compromise the sealability or integrity of the joint upon return to service. Some of the damage modes to look for are as follows:

- Sealing surface damage from mechanical abuse or corrosive attack – slight roughening okay, but radial scratches must be polished out.

- Warpage from uneven or excessive bolt loading or from thermal excursions – warpage greater than 1/16 in (1.59 mm) across the face of the flange requires further evaluation and/or repair.
- Backside radius cracking from excessive bolt loading, overpressure, raised face mating surfaces, or applied piping forces – superficial cracking is not uncommon, but pronounced cracking exposing the structural layers requires repair before returning to service.
- Gusset cracking from excessive bolt loading or applied piping forces – gussets are designed to act mostly in compression, so minor cracking or separation at the back of the flange face is considered acceptable, while other damage or cracking may not be.
- Radial cracking from excessive bolt loading and/or overpressure – radial cracking is not considered acceptable and generally will require flange replacement.

If any of the above forms of damage are found, it is highly recommended that an FRP specialist be contacted regarding evaluation of fitness for service. These forms of damage often indicate that other problems exist, such as improper gasketing, improper maintenance procedures, or even process upsets. If these root causes are not identified and addressed, an unplanned loss of containment can occur. An FRP specialist can assist with identification of the root causes, as well as the solutions.

Bolt/Fastener Inspection

During reassembly, inspection of the fasteners is also an important step towards ensuring a long-lived and reliable joint seal. The threads of the nuts, bolts, and/or studs should be inspected for corrosion, galling, and mechanical damage. A tap or die can be used to clean up any minor imperfections. Since the condition of the fastener plays a large role in how well the applied bolt torque is converted into clamping force, fasteners should be replaced if they exhibit signs of damage. Test fitting the nuts on the bolts or studs is a good way to double-check the condition of the fasteners. Any damage that effectively increases the turning friction of the nut on the bolt or stud will reduce the amount of clamping force generated for a given torque, and this translates into a lower compressive stress on the gasket. As spelled out in an earlier section of this chapter, application of the proper thread lubricant is essential prior to assembly of the fasteners.

TM Registered Trademark - Radiator Specialty Company

References

1. Sun, L., "Bolted Flanged Connections Made of Fiber Reinforced Plastic Materials," Ph.D. Dissertation, 1995, Concordia University, Montreal, Canada.

2. National Bureau of Standards, Voluntary Product Standard PS 15-69, *Custom Contact-Molded Reinforced Polyester Chemical-Resistant Process Equipment,* 1969.

3. Shigley, J.E., Mischke, C.R., *Standard Handbook of Machine Design, 2nd Edition,* Chapter 25 "Lubrication", ©1996 McGraw-Hill.

4. Private communication, Bill Bittner Sr., Bittner Industries.

5. Bickford, John H., *An Introduction to the Design and Behavior of Bolted Joints, 3rd Edition,* Chapter 7, ©1995 Marcel Dekker, ISBN 0-8247-9297-1.

6. Estrada, H. and Parsons, I.D., "Design Procedures for an FRP Stub-Flange Joint," Journal of Pressure Vessel Technology, Vol.124, Nov. 2002, pp. 397-404.

7. Private communication, Juan Bustillos, The Dow Chemical Company.

8. MTI Publication No. 14, Manual on Closure Bolting for The Chemical Process Industries.

Appendix A

Example Flange Calculation:

ASME RTP-1 Method #1

This Appendix presents the step-by-step procedure for designing an example flange using Method #1 (Appendix NM-2) of ASME RTP-1.

Nomenclature

A = outside diameter of flange (in)

A_B = cross-sectional area of bolts using the root diameter of the thread or diameter of unthreaded portion, whichever is less (in^2)

A_m = total required cross-sectional area of bolts (in^2)

a = nominal bolt diameter (in)

B = inside diameter of flange (in)

b = effective gasket or joint-contact surface seating width (in)

C = bolt circle diameter (in)

d = shape factor

d_1 = bolt hole diameter (in)

e = shape factor

F = shape factor (from Fig. NM2-3)

f = hub stress correction factor (from Fig. NM2-5)

G = diameter at location of gasket load reaction (in)

g_0 = thickness of hub at small end (in)

g_1 = thickness of hub at back of flange (in)

H = total hydrostatic end force (lb)

H_D = hydrostatic end force on area inside of flange (lb)

H_G = gasket load (lb)
H_{gy}' = compression load required to seat gasket outside G diameter (lb)
H_P = total joint-contact-surface compression load (lb)
H_P' = total adjusted joint-contact-surface compression for full-face gasket flange (lb)
H_T = difference between total hydrostatic end force and the hydrostatic end force area inside the flange (lb)
h = hub length (in)
h_D = radial distance from bolt circle to the circle on which H_D acts. (in)
h_G = circle on which H_G acts (in)
h_G' = radial distance from bolt circle to the circle on which H_G acts (in)
h_G'' = lever arm
h_T = radial distance from bolt circle to the circle on which H_T acts (in)
h_0 = shape factor (in)
K = ratio of outside diameter of flange to inside diameter of flange
L = shape factor
M = unit load, operating (lb)
M_D = component of moment due to H_D (in-lb)
M_G = component of moment due to H_G operating (in-lb)
M_O = total moment acting upon flange, operating conditions (in-lb)
M_T = component of moment due to H_T (in-lb)
m = gasket factor
N_1 = shape factor
n = number of bolts
P = design pressure (psi)
R = radial distance from bolt circle to point of intersection of hub and back of flange (in)
S_a = allowable bolt stress at atmospheric temperature (psi)
S_b = allowable bolt stress at design temperature (psi)
S_{fo} = allowable design stress for flange material at design temperature (psi)
S_H = calculated longitudinal stress in the hub (psi)
S_R = calculated radial stress in flange (psi)
S_{RAD} = calculated radial stress at bolt circle due to reverse moment (psi)
S_T = calculated tangential stress in flange (psi)
T = shape factor
t = flange thickness
U = shape factor
V = shape factor
W = average flange bolt load (lb)
W_{m_1} = required bolt load, operating conditions (lb)
W_{m_2} = required bolt load, gasket seating (lb)
Y = shape factor
y = gasket or joint-contact surface unit seating load (psi)
Z = shape factor

Flange Specifications and Initial Design Parameters

The example flange has the following specifications:

- A full-faced flange with inside diameter B = 24.0 in (61 cm)
- Design pressure P = 25 psi (17 MPa)
- A gasket with $m = 0.5$ and $y = 150$ psi (1 MPa)
- Bolt allowable tensile stress $S_a = S_b = 18{,}800$ psi (129 MPa)
- Class II laminate allowable design stress $S_{fo} = 3{,}000$ psi (21 MPa)

In addition, the following parameters are used in the design based on those recommended in Table NM2-1 and Table NM2-3 of RTP-1:

- Flange face diameter A = 32 in (81 cm)
- Bolt circle diameter C = 29.5 in (75 cm)
- Number of bolts n = 20
- Bolt diameter a = 1.25 in (3 cm)
- Bolt hole diameter $d_1 = 1.375$ in (3 cm)
- Hub thickness at small end $g_0 = 0.313$ in (8 mm)
- Initial guess for flange thickness t = 1.60 in (4 cm)
- Hub length h = 3t = 4.80 in (12 cm)

Note that this procedure is iterative. That is, the designer must make an initial guess for the flange thickness. The stresses in the flange are then calculated using this thickness, and if any of these stresses exceed the laminate allowable design stress, then the designer must choose another thickness and perform the calculations again.

Flange Design Calculations

Required Bolt Load

The following geometric parameters must be calculated to determine the required bolt load during gasket seating and under operating conditions (see Figure A1):

$$h_G = \frac{(C-B)}{4}$$

$$= \frac{(29.5-24.0)}{4} = 1.38 \text{ in}$$

$$h_G^{'} = \frac{(A-C)}{4}$$

$$= \frac{(32-29.5)}{4} = 0.63 \text{ in}$$

$$G = C - 2h_G$$

$$= 29.5 - 2(1.38) = 26.7 \text{ in}$$

$$b = 0.25(C - B)$$

$$= 0.25(29.5 - 24) = 1.38 \text{ in}$$

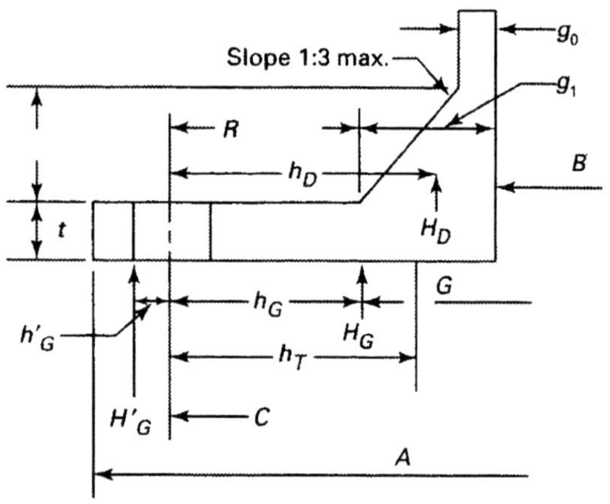

Source: ASME RTP-1-2000 , p.211

Figure A1. Flange Geometric Parameters and Forces.

The bolt load required to seat the gasket is given by

$$Wm_2 = b\pi Gy + H_{gy}^{'}$$

where

$$H_{gy}^{'} = \frac{h_G}{h_G^{'}}(b\pi Gy)$$

Thus

$$Wm_2 = (1.38)\pi(26.7)(150) + \left(\frac{1.38}{0.63}\right)\left[(1.38)\pi(26.7)(150)\right] = 55,465 \text{ lb}$$

or 2,773 lb per bolt.

The bolt load required under operating conditions is given by

$$Wm_1 = H + H_P + H_P^{'}$$

where H is the hydrostatic end force given by

$$H = 0.7854G^2P$$
$$= 0.7854(26.7)^2(25) = 14,050 \text{ lb}$$

and Hp and Hp' are contact surface compression forces given by

$$H_P = 2b\pi GmP$$
$$= 2(1.38)\pi(26.7)(0.5)(25) = 2,889 \text{ lb}$$

$$H_P^{'} = \left(\frac{h_G}{h_G^{'}}\right)H_P$$
$$= \left(\frac{1.38}{0.63}\right)(2,889) = 6,355 \text{ lb}$$

Thus

$Wm_1 = 14,050 + 2,889 + 6,355 = 23,294 \text{ lb}$ or 1,165 lb per bolt.

The total bolt area A_m is given by the greater of Wm_2/S_a or Wm_1/S_b

$$\frac{Wm_1}{S_b} = \frac{23,394}{18,800} = 1.24 \text{ in}^2$$

$$\frac{Wm_2}{S_a} = \frac{55,465}{18,800} = 2.95 \text{ in}^2$$

Hence, A_m = 2.95 in². Note that for the example flange, the bolt root area is given by

$$A_b = 0.25\pi(0.91a - 0.055)^2$$
$$= 0.25\pi[0.91(1.25) - 0.055]^2 = 0.92 \text{ in}^2$$

The average flange bolt load is then

$$W = 0.5(A_m + nA_b)S_a$$
$$= 0.5[2.95 + 20(0.92)](18{,}800) = 200{,}692 \text{ lb}$$ or 10,035 lb per bolt.

Moments on the Flange

To determine the stresses in the flange and hub, the moments acting on the flange must be calculated as follows (see Figure A1):

$$g_1 = \frac{t}{2} + g_0$$
$$= \frac{1.60}{2} + 0.313 = 1.11 \text{ in}$$

$$H_D = 0.7854B^2P$$
$$= 0.7854(24.0)^2(25) = 11{,}310 \text{ lb}$$

$$h_D = R + 0.5g_1$$
$$= 1.64 + 0.5(1.11) = 2.19 \text{ in}$$

$$M_D = H_D h_D$$
$$= (11{,}310)(2.19) = 24{,}808 \text{ in-lb}$$

$$H_T = H - H_D$$
$$= 14{,}050 - 11{,}310 = 2{,}740 \text{ lb}$$

$$h_T = 0.5(R + g_1 + h_G)$$
$$= 0.5(1.64 + 1.11 + 1.38) = 2.06 \text{ in}$$

$$M_T = H_T h_T$$
$$= (2{,}740)(2.06) = 5{,}652 \text{ in-lb}$$

$$M_O = M_D + M_T$$
$$= 24,808 + 5,652 = 30,460 \text{ in-lb}$$

$$M = \frac{M_O}{B}$$

$$H_G = W - H$$
$$= 200,692 - 14,050 = 186,642 \text{ lb}$$

$$h_G^{"} = \frac{(h_G h_G^{'})}{(h_G + h_G^{'})}$$
$$= \frac{(1.38)(0.63)}{(1.38 + 0.63)} = 0.43 \text{ in}$$

$$M_G = H_G h_G^{"}$$
$$= (186,642)(0.43) = 80,256 \text{ in-lbs}$$

Shape Factors

The formulae for calculating stresses in the flange from the moments incorporate various shape factors. These shape factors are determined as follows:

$$K = \frac{A}{B} = \frac{32}{24} = 1.33$$

$$\frac{g_1}{g_0} = \frac{1.11}{0.313} = 3.55$$

$$h_0 = (Bg_0)^{1/2} = [(24)(0.313)]^{1/2} = 2.74$$

$$\frac{h}{h_0} = \frac{4.80}{2.74} = 1.75$$

$T = 1.78$ (from Figure NM2-2)

$Z = 3.57$ (from Figure NM2-2)

$Y = 6.91$ (from Figure NM2-2)

$U = 7.59$ (from Figure NM2-2)

$F = 0.57$ (from Figure NM2-3)

$V = 0.04$ (from Figure NM2-4)

$f = 1.0$ (from Figure NM2-5)

$$e = \frac{F}{h_0} = \frac{0.57}{2.74} = 0.21$$

$$d = \left(\frac{U}{V}\right) h_0 g_0^2 = \left(\frac{7.59}{0.04}\right)(2.74)(0.313)^2 = 50.9$$

$$N_1 = \frac{4}{3} te + 1 = \frac{4}{3}(1.60)(0.21) + 1 = 1.44$$

$$L = \frac{(te+1)}{T} + \frac{t^3}{d} = \frac{[(1.60)(0.21)+1]}{1.78} + \frac{(1.60)^3}{50.9} = 0.83$$

Stress Calculations

The following stresses are calculated and checked to ensure that they do not exceed the allowable design stress for the Class II laminate S_{fo} = 3,000 psi:

(a) Longitudinal Stress in the Hub

$$S_H = \frac{fM}{Lg_1^2}$$

$$= \frac{(1.0)(1269)}{(0.83)(1.11)^2} = 1,241 \text{ psi} < 3,000 \text{ psi}$$

(b) Radial Stress in the Flange

$$S_R = \frac{N_1 M}{Lt^2}$$

$$= \frac{(1.44)(1269)}{(0.83)(1.60)^2} = 860 \text{ psi} < 3,000 \text{ psi}$$

(c) Tangential Stress in Flange

$$S_T = \frac{YM}{t^2} - ZS_R$$

$$= \frac{(6.91)(1269)}{(1.60)^2} - (3.57)(860) = 355 \text{ psi} < 3{,}000 \text{ psi}$$

(d) Greater of $0.5(S_H + S_R)$ and $0.5(S_H + S_T)$

$$0.5(1241 + 860) = 1{,}051 \text{ psi}$$

$$0.5(1241 + 355) = 798 \text{ psi}$$

$$1{,}051 \text{ psi} < 3{,}000 \text{ psi}$$

(e) Radial Stress at the Bolt Circle

$$S_{RAD} = \frac{6M_G}{t^2(\pi C - nd_1)}$$

$$= \frac{6(80256)}{(1.60)^2\left[(\pi)(29.5) - (20)(1.375)\right]} = 2{,}886 \text{ psi} < 3{,}000 \text{ psi}$$

Because all of these stresses are less than 3,000 psi (21 MPa), the chosen flange thickness t = 1.60 in (4 cm) is appropriate. If one or more of these stresses exceeded 3,000 psi (21 MPa), a greater thickness value would be chosen and the calculations would be performed again. This process would be repeated until an appropriate thickness was found.

Appendix B

Example Flange Calculation:

ASME RTP-1 Method #2

This Appendix presents the step-by-step procedure for designing an example flange using Method #2 (Appendix NM-12) of ASME RTP-1.

Nomenclature

A_b = bolt root area (in^2)
B.C.D. = bolt circle diameter (in)
b_s = bolt spacing (in)
D_b = diameter of bolts (in)
e_c = edge clearance (in)
h_c = hub clearance (in)
h_h = hub height (in)
h_t = hub thickness (in)
I.D. = flange inner diameter (in)
m = gasket factor
N = number of bolts in flange
n_t = nozzle wall thickness
O.D. = flange outside diameter (in)
P = internal design pressure (psi)
P_a = axial load on flange (lb)
T_{max} = maximum bolt torque (ft-lb)
T_{min} = minimum bolt torque (ft-lb)
t_g = flange thickness required to seat gasket (in)
t_o = flange thickness under operating loads (in)

y	=	gasket seating stress (psi)
σ_b	=	design bolt tensile stress (psi)
σ_{bf}	=	bolt stress at flange design stress (psi)
σ_{bg}	=	bolt stress required to seat gasket (psi)
σ_{bo}	=	bolt stress under operating loads (psi)
σ_f	=	design flange flexural stress (psi)
σ_n	=	design nozzle wall tensile stress (psi)

Flange Specifications and Initial Design Parameters

The example flange has the following specifications:

- A full-faced flange with inside diameter I.D. = 24.0 in (61 cm)
- Design pressure P = 25.0 psi (.17 MPa)
- A gasket with $m = 0.5$ and $y = 150.0$ psi (1 MPa)
- Design bolt tensile stress σ_b = 18,800 psi (129 MPa)
- Class II laminate design flexural stress σ_f = 3,000 psi (21 MPa)
- Class II laminate design nozzle wall tensile stress σ_n = 3,000 psi

In addition, the following parameters are used in the design based on Class 150 ANSI flanges:

- Flange face diameter O.D. = 32.0 in (81 cm)
- Bolt circle diameter B.C.D. = 29.5 in (75 cm)
- Number of bolts N = 20
- Bolt diameter D_b = 1.25 in (3 cm)

Flange Design Calculations

Nozzle Wall Thickness

The thickness of the nozzle wall (see Figure B1) must be known to calculate the bending moments in the flange due to the gasket seating and operational forces. The thickness of the nozzle wall necessary to resist internal pressure and axial load is calculated as follows:

The nozzle wall thickness necessary to ensure that the stress due to internal pressure is lower than the design nozzle wall tensile stress is

$$n_p = P\frac{I.D.}{2\sigma_n}$$

$$= (25)\frac{(24)}{2(3000)} = 0.10 \text{ in}$$

The nozzle wall thickness necessary to ensure that the stress due to axial loading is lower than the design nozzle wall tensile stress is

$$n_p = \frac{P_a}{\pi I.D.\sigma_n}$$

$$= \frac{0}{\pi(24)(3000)} = 0$$

From the above results, the nozzle wall thickness n_t = 0.10 in (2 mm).

Other Flange Geometric Parameters

Other geometric parameters that are required to calculate the bending stresses in the flange when seating the gasket and under operating conditions are as follows (see Figure B1):

Bolt root area:

$$A_b = 0.25\pi(0.91D_b - 0.055)^2$$

$$= 0.25\pi[0.91(1.25) - 0.055]^2 = 0.92 \text{ in}^2$$

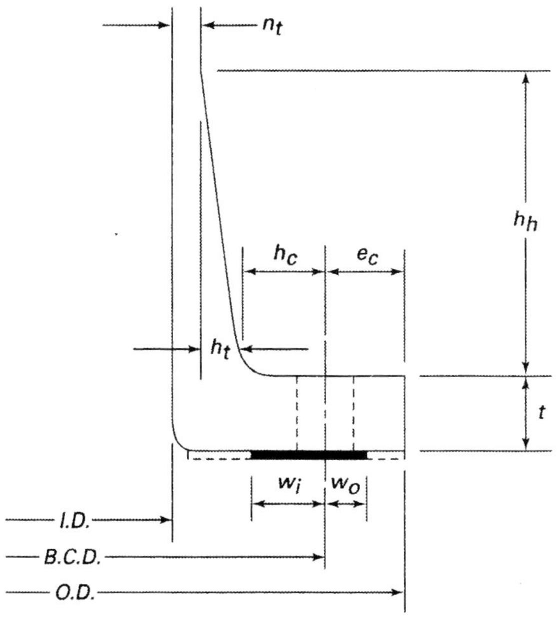

Source: ASME RTP-1-2000 , p.302

Figure B1. Flange Geometric Parameters.

Effective gasket width inside the bolt circle diameter:

$$w_i = 0.25(B.C.D. - I.D.)$$

$$= 0.25(29.5 - 24.0) = 1.38 \text{ in}$$

Effective gasket width outside the bolt circle diameter:

$$w_o = 0.25(O.D. - B.C.D.)$$

$$= 0.25(32.0 - 29.5) = 0.63 \text{ in}$$

Moment arm for axial load:

$$a_1 = 0.5(B.C.D. - I.D. - n_t)$$

$$= 0.5(29.5 - 24.0 - 0.10) = 2.70 \text{ in}$$

Moment arm for pressure load:

$$a_2 = 1.5w_i$$

$$= 1.5(1.38) = 2.07 \text{ in}$$

Moment arm for gasket load:

$$a_3 = 0.5(w_o + w_i) - w_o$$

$$= 0.5(0.63 + 1.38) - 0.63 = 0.38 \text{ in}$$

Length of inside effective gasket:

$$L_i = \pi(B.C.D. - w_i)$$

Length of outside effective gasket:

$$L_o = \pi(B.C.D. + w_o)$$
$$= \pi(29.5 + 0.63) = 94.7\text{in.}$$

Effective bolt circle length:

$$L_e = \pi B.C.D. - N(D_b + 0.125)$$

$$= \pi(29.5) - 20(1.25 + 0.125) = 65.2 \text{ in}$$

Forces on the Flange

The forces on the flange for gasket seating and under operating conditions are calculated as follows (see Figure B2):

Bolt loading during gasket seating (per bolt):

$$F_B = \frac{(w_o L_o + w_i L_i)y}{N}$$

$$= \frac{[(0.63)(94.7) + (1.38)(88.3)](150)}{20} = 1{,}361 \text{ lb}$$

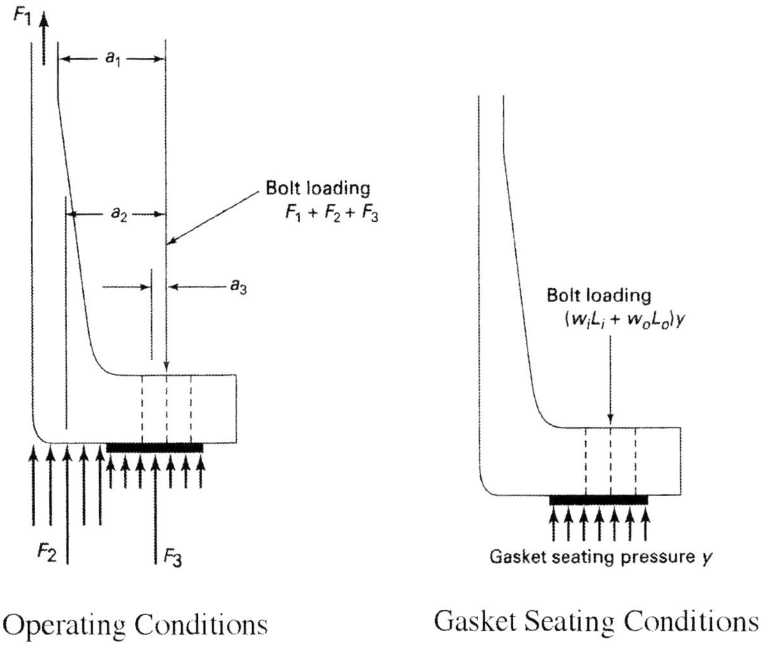

Source: ASME RTP-1-2000, p.303

Figure B2. Forces on the Flange.

Axial load on flange during operation:

$$F_1 = P_a + 0.25\pi(I.D.)^2 P$$

$$= 0 + 0.25\pi(24.0)^2(25) = 11{,}310 \text{ lb}$$

Pressure load inside gasket during operation:

$$F_2 = \pi w_i (I.D. + w_i) P$$

$$= \pi(1.38)(24.0 + 1.38)(25) = 2{,}751 \text{ lb}$$

Load from gasket compression during operation:

$$F_3 = (w_o L_o + w_i L_i) mP$$

$$= [(0.63)(94.7) + (1.38)(88.3)](0.5)(25) = 2{,}269 \text{ lb}$$

Gasket Seating Conditions

The stress in each bolt during gasket seating is

$$\sigma_{bg} = \frac{(w_o L_o + w_i L_i) y}{N A_b}$$

$$= \frac{[(0.63)(94.7) + (1.38)(88.3)](150)}{(20)(0.92)} = 1{,}480 \text{ psi}$$

The thickness of the flange necessary to ensure that the flange does not exceed its design flexural stress during gasket seating is

$$t_g = \sqrt{\frac{3y}{L_e \sigma_f}(w_o^2 L_o + w_i^2 L_i)}$$

$$= \sqrt{\frac{3(150)}{(65.2)(3000)}[(0.63)^2(94.7) + (1.38)^2(88.3)]} = 0.69 \text{ in}$$

Flange Operating Conditions

The stress in each bolt during flange operation is

$$\sigma_{bo} = \frac{F_1 + F_2 + F_3}{N A_b}$$

$$= \frac{11310 + 2751 + 2269}{(20)(0.92)} = 888 \text{ psi}$$

The thickness of the flange necessary to ensure that the flange does not exceed its design flexural stress during operation is

$$t_o = \sqrt{\frac{6}{L_e \sigma_f}(F_1 a_1 + F_2 a_2 + F_3 a_3)}$$

$$= \sqrt{\frac{6}{(65.2)(3000)}[(11310)(2.70) + (2751)(2.07) + (2269)(0.38)]} = 1.07 \text{ in}$$

Flange Thickness

The final flange thickness is the greater of the thickness required for gasket seating (t_g) and the thickness required for operational loading (t_o). In this case, the operational loading determines the flange thickness:

t = 1.07 in (3 cm)

Note also that for both the gasket seating and operational conditions, the stress in each bolt is less than the design tensile stress of the bolts, as required.

Bolt Torque

The minimum bolt torque is determined by considering the bolt stress required for gasket seating and under operating conditions:

$$T_{\min} = \frac{\sigma_{bm} A_b D_b}{60} \quad \text{(in ft-lb)}$$

where σ_{bm} is the larger of σ_{bg} and σ_{bo}. For the example flange, $\sigma_{bm} = \sigma_{bg} = 1{,}480$ lbs. The minimum bolt torque is then

$$T_{\min} = \frac{(1480)(0.92)(1.25)}{60} = 28.4 \text{ ft-lb}$$

The maximum bolt torque is determined by considering the maximum bolt stress the flange and the bolt can withstand without exceeding their respective design stress limits. The maximum torque that can be applied without the flange exceeding its design stress is

$$T_f = \frac{\sigma_{bf} A_b D_b}{60}$$

where σ_{bf} is

$$\sigma_{bf} = \frac{L_e t^2 \sigma_f (w_o L_o + w_i L_i)}{2NA_b (w_o^2 L_o + w_i^2 L_i)}$$

$$= \frac{(65.2)(1.07)^2(3000)[(0.63)(94.7) + (1.38)(88.3)]}{2(20)(0.92)[(0.63)^2(94.7) + (1.38)^2(88.3)]} = 333{,}587 \text{ psi}$$

Thus

$$T_f = \frac{(333587)(0.92)(1.25)}{60} = 6{,}394 \text{ ft-lb}$$

The maximum torque that can be applied to the bolts without exceeding the design bolt tensile stress is

$$T_b = \frac{\sigma_b A_b D_b}{60}$$

$$= \frac{(18800)(0.92)(1.25)}{60} = 360.3 \text{ ft-lb}$$

Therefore, the maximum torque is determined by the design bolt tensile stress and is

$$T_{max} = 360.3 \text{ ft-lb}$$

Hub Thickness and Height

Minimum hub thickness (see Figure B1):

$$h_t = 0.5t$$

$$= 0.5(1.07) = 0.54 \text{ in}$$

Minimum hub height (see Figure B1):

$$h_h = 6h_t$$

$$= 6(0.54) = 3.24 \text{ in}$$

Bolt Spacing and Clearances

Bolt spacing:

$$b_s = \frac{\pi B.C.D.}{N}$$

$$= \frac{\pi(29.5)}{20} = 4.63 \text{ in}$$

Maximum bolt spacing:

$$b_{s,\max} = 2D_b + 6t$$

$$= 2(1.25) + 6(1.07) = 8.92 \text{ in}$$

Edge clearance:

$$e_c = 2w_o$$

$$= 2(0.63) = 1.26 \text{ in}$$

Minimum edge clearance:

$$e_{c,\min} = 0.917D_b + 0.167$$

$$= 0.917(1.25) + 0.167 = 1.31 \text{ in}$$

Hub clearance:

$$h_c = 2w_i - n_t - h_t$$

$$= 2(1.38) - 0.10 - 0.54 = 2.12 \text{ in}$$

Minimum hub clearance:

$$h_{c,\min} = 1.125D_b + 0.3$$

$$= 1.125(1.25) + 0.3 = 1.71 \text{ in}$$

Appendix C

Example Flange Calculation:

Lap-Joint Flange and Ring

This Appendix presents the step-by-step procedure for designing an example flange using the proposed lap-joint flange design method:

Nomenclature

Ab	=	bolt root area (in^2)
BC	=	bolt circle diameter (in)
d	=	diameter of bolts (in)
hh	=	hub height (in)
ht	=	hub thickness (in)
IDf	=	flange inner diameter (in)
m	=	gasket factor
N	=	number of bolts in flange
tn	=	nozzle wall thickness
ODr	=	flange outside diameter (in)
P	=	internal design pressure (psi)
Pa	=	axial load on flange (lb)
Tmax	=	maximum bolt torque (ft-lb)
Tmin	=	minimum bolt torque (ft-lb)
tg	=	flange thickness required to seat gasket (in)
to	=	flange thickness under operating loads (in)
y	=	gasket seating stress (psi)
σbolt	=	design bolt tensile stress (psi)
σbf	=	bolt stress at flange design stress (psi)

σbg = bolt stress required to seat gasket (psi)
σbo = bolt stress under operating loads (psi)
σf = design flange flexural stress (psi)
σt = design nozzle wall tensile stress (psi)
tf = minimum required FRP flange thickness (in)
τflg = flange shear stress (psi)
IDr = ring inside diameter (in)
$\tau aflg$ = allowable flange shear stress (psi)
Efrp = nozzle flexural modulus of elasticity
tring = chosen backing ring thickness (in)
$\tau aring$ = allowable ring shear stress (psi)
E = ring tensile modulus of elasticity (psi)
$\tau ring$ = ring shear stress (psi)
ya = deflection at outer edge of ring (in)
θo = angular rotation of ring at outer edge (radians)
θi = angular rotation of ring at inner edge (radians)
ODf = outside diameter of the FRP stub (in)
σf = laminate design flexural stress (psi)
σn = laminate design tensile stress (psi)

Flange Specifications and Initial Design Parameters

The example flange has the following specifications:

- A lap-joint flange with inside diameter I.D. = 24.0 in (61 cm)
- Design pressure P = 25.0 psi (.17 MPa)
- A gasket with *m* = *0.5* and *y* = 50.0 psi (.34 MPa)
- Design bolt tensile stress σ_b = 18,800 psi (129 MPa)
- Laminate design flexural stress σ_f = 3,000 psi (21 MPa)
- Laminate design nozzle wall tensile stress σ_n = 3,000 psi (21 MPa)

In addition, the following parameters are used in the design based on those recommended in Table NM2-1 of the ASME RTP-1:

- Ring outside diameter OD = 29.5 in (75 cm)
- Bolt circle diameter BC = 28.0 in (71 cm)
- Number of bolts N = 20
- Bolt diameter d = 0.5 in (12 mm)

Flange Design Calculations

Nozzle Wall Thickness

The thickness of the nozzle wall (see Figure C1) must be known to calculate the bending moments in the flange due to the gasket seating and operational forces. The thickness of the nozzle wall necessary to resist internal pressure and axial load is calculated as follows:

The nozzle wall thickness necessary to ensure that the stress due to internal pressure is lower than the design nozzle wall tensile stress is

$$tp = \frac{P * IDf}{2 * \sigma n} \qquad tp = 0.10$$

The nozzle wall thickness necessary to ensure that the stress due to axial loading is lower than the design nozzle wall tensile stress is

$$ta = \frac{Pa}{\pi * IDf * \sigma n} \qquad ta = 0$$

Note: The required nozzle wall thickness n_t, is the larger of tp and ta.

From the above results, the calculated nozzle wall thickness n_t = 0.10 in (2 mm).

ASME RTP-1 paragraph 3-200 (f) states minimum thickness shall not be less than 0.22 inches thus:

$$tn = 0.22$$

Other Flange Geometric Parameters

Other geometric parameters that are required to calculate the bending stresses in the flange when seating the gasket and under operating conditions are as follows (see Figure C1):

Bolt root area (in^2):

$$Ab = 0.25 * \pi * (0.91 * d - 0.055 \wedge 2) \wedge 2 \qquad Ab = 0.126$$

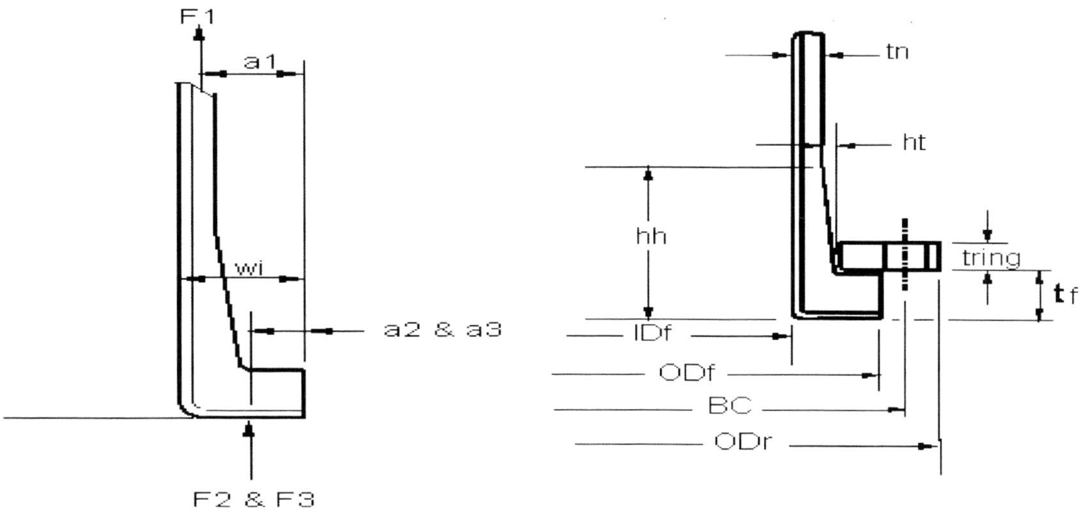

Figure C1. Flange Geometric Parameters.

Effective gasket width inside the bolt circle diameter:

$$wi = \frac{(ODf - IDf)}{2}$$

$$wi = 1.688$$

Moment arm for axial load:

$$a1 = wi - \frac{tn}{2}$$

$$a1 = 1.577$$

Moment arm for pressure load:

$$a2 = \frac{wi}{2}$$

$$a2 = 0.844$$

Moment arm for gasket load:

$$a3 = a2$$

$$a3 = 0.844$$

Length of inside effective gasket:

$$Li = \pi(ODf - wi)$$

$$Li = 80.7$$

Effective bolt circle length:

$$Le = \pi * ODf$$

$$Le = 86.001$$

Forces on the Flange

The forces on the flange for gasket seating and under operating conditions are calculated as follows (see Figure C2):

Bolt loading during gasket seating (per bolt):

$$Fb = \frac{wi * Li * y}{N}$$

$$Fb = 340.452$$

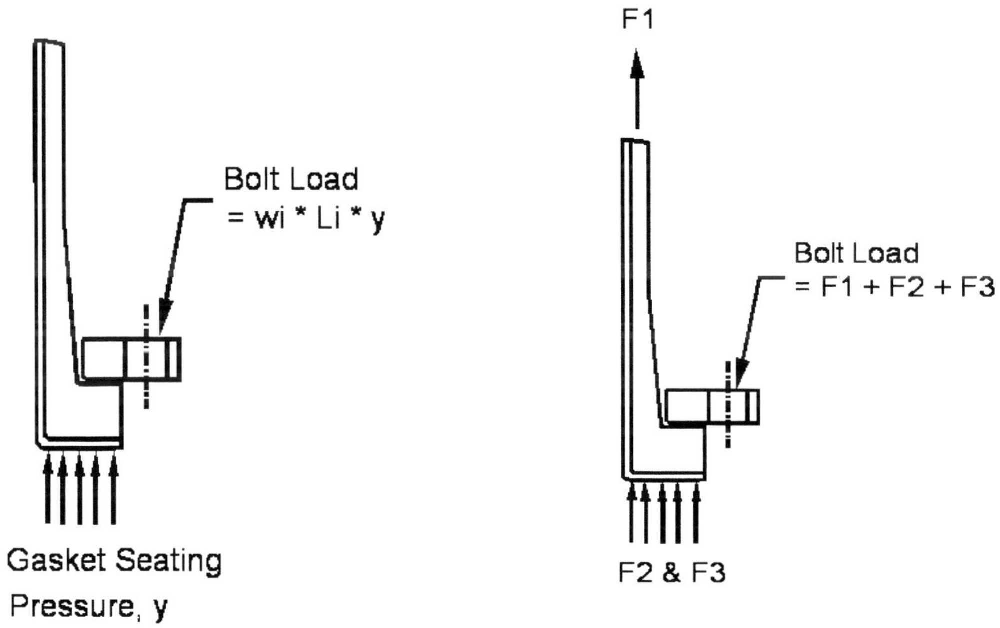

Figure C2. Forces on the Flange.

Axial load on flange during operation:

$$F1 = Pa + (0.25 * \pi * (IDf) \wedge 2 * P$$

$$F1 = 11310$$

Pressure load inside gasket during operation:

$$F2 = \pi * wi * (IDf + wi) * P$$

$$F2 = 3405$$

Load from gasket compression during operation:

$$F3 = wi * Li * m * P$$

$$F3 = 1702$$

Gasket Seating Conditions

The stress in each bolt during gasket seating is

$$\sigma bg = \frac{wi * Li * y}{N * Ab}$$

$$\sigma bg = 2709.229$$

The thickness of the flange necessary to ensure that the flange does not exceed its design flexural stress during gasket seating is

$$tg = \sqrt{\frac{3 * y * (wi)^\wedge 2 * Li}{Le * \sigma f}}$$

$$tg = 0.427$$

Flange Operating Conditions

The stress in each bolt during flange operation is

$$\sigma bo = \frac{F1 + F2 + F3}{N * Ab}$$

$$\sigma bo = 6532$$

The thickness of the flange necessary to ensure that the flange does not exceed its design flexural stress during operation is

$$to = \sqrt{\frac{6 * (F1 * a1 + (F2 + F3) * a2}{Le * \sigma f}}$$

$$to = 0.838$$

Flange Thickness

The final flange thickness is the greater of the thickness required for gasket seating (tg) and the thickness required for operational loading (to). In this case, the operational loading determines the flange thickness:

$$tf = 0.838$$

Note also that for both the gasket seating and operational conditions, the stress in each bolt is less than the design tensile stress of the bolts, as required.

Bolt Torque

The minimum bolt torque is determined by considering the bolt stress required for gasket seating and under operating conditions:

$$T\min = \frac{\sigma bm * Ab * d}{60} \text{ ft-lb}$$

where σbm is the larger of σbg and σbo. The minimum bolt torque is then

$T\min = 6.8$

The maximum bolt torque is determined by considering the maximum bolt stress the flange and the bolt can withstand without exceeding their respective design stress limits. The maximum torque that can be applied without the flange exceeding its design stress is

$$T\max = \frac{\sigma bf * Ab * d}{60} \text{ (ft-lb)}$$

where σbf is

$$\sigma bf = \frac{Le * tf \wedge 2 * \sigma f * wi}{2 * N * Ab * wi \wedge 2}$$

$$\sigma bf = 10450$$

Thus

$$T\max = 10.9$$

The maximum torque that can be applied to the bolts without exceeding the design bolt tensile stress is

$$Tb = \frac{\sigma b * Ab * d}{60}$$

$$Tb = 20.9$$

Therefore, the maximum torque is determined by the design bolt tensile stress and is

$$T\max = 10.9$$

Hub Thickness and Height

Minimum hub thickness (see Figure C)

$$ht = \frac{tf}{2}$$

$$ht = 0.419$$

Minimum hub height (see Figure C1)

$$hh = 4 * tf$$

$$hh = 3.352$$

Lap Joint Ring Design

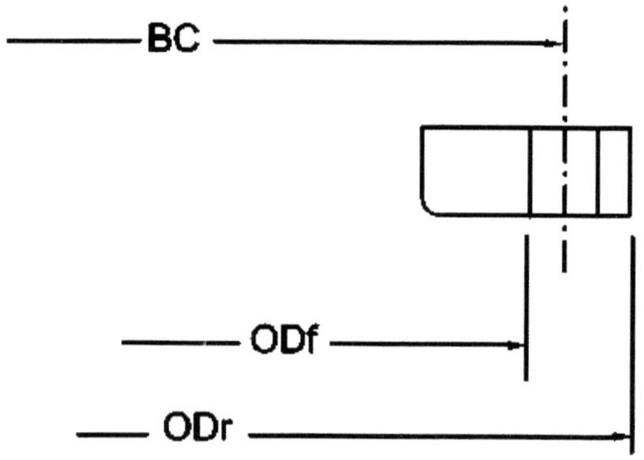

Figure C3. Ring Dimensions.

The analysis for the ring follows that presented in *Formulas for Stress and Strain* by Roark (5th edition Table 24 Case 1k) additional constants and nomenclature (v = Poisson's ratio):

$$a = \frac{ODr}{2} \qquad b = \frac{ODf}{2} \qquad v = 0.3 \qquad ro = b + \frac{d + 0.063}{2}$$

$$D = \frac{E * tring \wedge 3}{12 * (1 - v \wedge 2)} \qquad D = 226200$$

The greater of:

$$w := \frac{wi \cdot Liy}{\pi \cdot ID_r} \quad \text{or} \quad w := \frac{F1 \cdot F3}{\pi \cdot ID_r}$$

$$C1 = \left(\frac{1+v}{2}\right)\left(\frac{b}{a}\ln\left(\frac{a}{b}\right)\right) + \left(\frac{1-v}{4}\right)\left(\frac{a}{b} - \frac{b}{a}\right) \qquad C1 = 0.071$$

$$C3 = \frac{b}{4*a}\left[\left(\left(\frac{b}{a}\right)^{\wedge}2+1\right) * \text{In}\left(\frac{b}{a}\right) + \left(\frac{b}{a}\right)^{\wedge}2-1\right] \qquad C3 = 0.00006$$

$$C4 = \frac{1}{2} * \left[(1+v)*\frac{b}{a} + (1-v)*\frac{a}{b}\right] \qquad C4 = 0.98$$

$$C6 = \frac{b}{4*a} * \left[\left(\frac{b}{a}\right)^{\wedge}2 - 1 + 2*\ln\left(\frac{a}{b}\right)\right] \qquad C6 = 0.00247$$

$$C7 = \frac{1}{2} * (1 - v^{\wedge}2) * \left(\frac{a}{b} - \frac{b}{a}\right) \qquad C7 = 0.068$$

$$C9 = \frac{b}{a} * \left[\left(\frac{1+v}{2}\right) * \ln\left(\frac{a}{b}\right) + \left(\frac{1-v}{4}\right) * \left[1 - \left(\frac{b}{a}\right)^{\wedge}2\right]\right] \qquad C9 = 0.068$$

$$L3 = \frac{ro}{4*a} * \left[\left[\left(\frac{ro}{a}\right)^{\wedge}2+1\right] * \ln\left(\frac{a}{ro}\right) + \left(\frac{ro}{a}\right)^{\wedge}2-1\right] \qquad L3 = 0.0000241$$

$$L6 = \frac{ro}{4*a} * \left[\left(\frac{ro}{a}\right)^{\wedge}2 - 1 + 2*\ln\left(\frac{a}{ro}\right)\right] \qquad L6 = 0.001352$$

$$L9 = \frac{ro}{a} * \left[\left(\frac{1+v}{2}\right) * \ln\left(\frac{a}{ro}\right) + \left(\frac{1-v}{4}\right) * \left[1 - \left(\frac{ro}{a}\right)^{\wedge}2\right]\right] \qquad L9 = 0.051$$

rotation at the inner edge of the ring (radians):

$$\theta b = \frac{-w*a^{\wedge}2}{D*C7} * \left(\frac{ro*C9}{b} - L9\right) \qquad \theta b = -0.054$$

rotation in degrees:

$$\theta i = \theta b * \left(\frac{180}{\pi}\right) \qquad \theta i = -3.093$$

rotation at the outer edge of the ring (radians):

$$\theta a = \frac{-w * a \wedge 2}{D} * \left[\frac{C4}{C7} * \left(\frac{ro * C9}{b} - L9 \right) - \frac{ro * C6}{b} + L6 \right] \qquad \theta a = -0.053$$

rotation in degrees:

$$\theta o = \theta a * \left(\frac{180}{\pi} \right) \qquad \theta o = -3.019$$

deflection at the outer edge (inches):

$$ya = \frac{-w * a \wedge 3}{D} * \left[\frac{C1}{C7} * \left(\frac{ro * C9}{b} - L9 \right) - \frac{ro * C3}{b} + L3 \right] \qquad ya = -0.057$$

shear stress on the ring (psi):

shear load (lbf): $Qb = \frac{w * ro}{b}$ $\qquad Qb = 211.31$

shear stress: $\tau ring = \frac{Qb}{tring}$ $\qquad \tau ring = 480.245$

seating width of the ring on the stub flange (inches):

$$width = \frac{ODf - IDr}{2} \qquad width = 1.068$$

hub stresses following Roark's Formulas for Stress and Strain 4th edition Table XIII case 15:

$$Mf\,\lg = w * \left[\frac{width}{2} \right] \qquad Mf\,\lg = 110.611$$

$$\lambda hub = \left(\frac{3 * (1 - v \wedge 2)}{\left(\frac{IDf}{2} \right) \wedge 2 * (tn + ht) \wedge 2} \right)^{0.25} \qquad \lambda hub = 0.462$$

$$\sigma hub = \frac{2 * Mf\,\lg}{(tn + ht)} * \lambda hub \wedge 2 * \frac{IDf}{2} \qquad \sigma hub = 885.682$$